Preface

There can be no doubt of the impact of electronics on our lives. It has made possible dramatic improvements in communication, information storage and retrieval, manufacturing techniques, and help for people with disabilities, to name just a few. Logic systems in general and computers in particular carry out complex and often repetitive tasks tirelessly. Fewer moving parts make systems more reliable. With suitable transducers the value of any measurable physical quantity can be converted into an electrical signal. This enables electronic monitoring and control of processes with a greatly increased speed and accuracy over other methods.

The unease felt by many when it comes to understanding electronics is due partly to the inherently silent and invisible operation of the electronic system. Unlike mechanical systems, it is not generally possible to 'take the lid off' and observe the process in action. Nevertheless it is possible to break the system down into smaller functional blocks whose behaviour can be more readily understood. We can then quickly move on from 'how does it work?' to the more exciting question of 'what can we use it for?'.

This approach has been adopted throughout this book. The main functional blocks are considered in sections 2, 3, 5 and 6. As well as examining their behaviour, much attention is given to how and where they are used. This is taken further in sections 4, 7 and 8 where three important applications are considered in greater detail. The section on components has deliberately been left to the end. It is envisaged that this be used as a reference rather than a starting point.

The student will wish to construct as many circuits as possible and usually component values are given in the circuit diagrams. In most cases the exact values are not critical. The power supply has generally been taken to be 9 V but most circuits described will work with supplies between 3 V and 15 V. The only exception to this is in digital circuits where a +5 V supply is a generally accepted norm and indeed essential for TTL circuits. More detailed practical investigation is fully covered in the associated blackline master pack **Systematic Electronics: Practical Investigations**.

It is hoped that through this book, the reader will gain a broad understanding of both the principles and applications of electronics.

Many people have contributed in some way to this book and I

would like to thank my colleagues in general and David Grace in particular for their valuable suggestions during the draft stage. Nevertheless any errors and opinions expressed herein are solely my own. I must especially thank Joan Angelbeck and Katie Sparkes at Edward Arnold for their patience and guidance and Donna Thynne for her tireless assistance with the photographs.

Thanks must also go to my pupils, both past and present, whose comments and questions have, however unwittingly, often caused me to think afresh and have been a frequent source of inspiration.

I am also very grateful to my wife for her tolerance and assistance with the preparation of the manuscript, my children for accepting my preoccupations over the past 18 months and finally my parents; not only for their help and encouragement throughout this project but also for initiating my earliest interest in electronics.

Peter G. Vaughan
London 1987

Acknowledgements

The publishers would like to thank the Southern Examining Group for permission to include questions from past papers, and Heinemann Educational Books Ltd for permission to use a figure from A. F. Abbott: *Ordinary Level Physics*.

The publisher's thanks are also due to the following for permission to reproduce copyright photographs:

BBC Hulton Picture Library: pp 1 left (Bettman Archive) & 52 left; Akai (UK) Ltd: p 1 right; Popperfoto: p 2; Maplin Electronic Supplies Ltd: p 3 top & circuit diagram; Ferranti Computer Systems Ltd: pp 6 top & 169; Barnaby's Picture Library: pp 37, 82 & 147 top left; CEGB/Balfour Beatty: p 12 top; Steve Richards: pp 14 top, 19 top, 52 right, 84 right & 87; Diana Lanham: p 14 bottom left; London Fire Brigade: p 14 right; John Robinson: p 15 top; Shure Brothers Inc. USA: p 15 bottom left; Philips Electronics: p 16; Ford Motor Company Ltd: p 15 bottom right; Royce Thompson Electric Ltd: p 19 bottom; Chubb Alarms Ltd: pp 20 top & 31 top; Wilson Grimes Products: p 23; Shell UK Ltd: p 24; Pinniger Gosling: p 31 bottom left; Katie Sparkes: p 31 bottom right; RS Components Ltd: pp 35, 198, 201 (a) (c) (d) & 204; Stone Chance: p 42; Tobias Baeuerle und Söhne: p 46; Cincinnati Milacron UK Co: p 53; Sally–Anne Greville–Heygate: p 72; Philip Harris Ltd: pp 74 top left & right & 126 left; Educational Electronics: p 74 bottom; Heathrow Airport Ltd 1987/P & P F James Photography Ltd: p 78; Fostex/Bandive Ltd: pp 79 left & 91; The London Philharmonic Orchestra: p 81; John Hornby Skewes & Co Ltd: p 83; Stuart Chorley: p 48 left; Linn Products Ltd: p 88; Sony (UK) Ltd: pp 90 & 124; Marantz Audio UK Ltd/N & G Insight: pp 93 & 96 bottom & top; Tektronix UK Ltd: pp 95 bottom, 127 right & 129; WEM: p 97; Roland UK Ltd: pp 98, 99 & 101 right; Casio Electronics Ltd: p 101 left; Marconi Communications Systems: pp 106 bottom left & 126 right; AMI Health Care Ltd/Dr M J Goldsmith: p 106 top left; Cambridge Systems Technology Ltd: p 106 centre right; Telefocus/ British Telecom: pp 127 left, 147 bottom left, 149, 160 left & right, 161 bottom & 166 top; Chronar III Ltd: p 132 left; NASA: p 132 right; Vessa Ltd: p 137; CB Magazine: p 146 right; The Meteorological Office: p 146 left; The Open University: p 147 bottom right; British Nuclear Fuels plc: p 147 top right; Peter Vaughan: p 148 right; John Grooms Association for the Disabled: p 148 left; Oracle: p 159; The Stock Photobank, London: p 161 top; Commodore Business Machines: p 165 left; Martin West: p 166 bottom: Geoffe Goode; p 167 left; Hewlett Packard Limited: p 167 right; Mullard Southampton: p 168; Motorola/Hi Tech PR Ltd: p 195 right.

The remaining photographs appearing in this book were taken by Donna Thynne and are the copyright of Edward Arnold Ltd. The parts were supplied by Maplin Electronic Supplies Ltd and the publishers wish to thank them for their time and generous assistance.

Special thanks must go to all the manufacturers mentioned above for their patient and kind help with this project.

Contents

Section 1: Basic Principles

Section 2: Switching Systems

Section 3: Digital Systems

Section 4: Audio Systems

Section 5: Analogue Systems

Section 6: Power Supplies

Section 7: Communication Systems

Section 8: Programmable Systems

Section 9: Components

Section 1

Basic Principles

1
Electronic Systems

What is electronics?

People first experimented with electric current over 200 years ago and useful devices like light bulbs and electric motors have been in use for at least 100 years. There is nothing new about electricity!

Electricity is probably the most useful form of energy because it can be turned into so many other kinds of energy (like sound, heat or light). As well as this, electricity is useful because it can be **controlled**.

Electronics is all about **controlling** the way in which electricity flows round a circuit.

This book will help you to understand not only how electric currents can be controlled and how electronic devices work but also how you can design and build your own devices and systems.

What is an electronic system?

In electronics, a **system** is made up from individual **circuits**. These circuits are combined together so that they can do something useful. Eventually the system produces an **output**. For example, a hi-fi system is designed to produce sound.

Edison in his library with the statue of 'Electricity'.

A hi-fi system is made up from several building blocks.

Block diagrams

The hi-fi system is made up from separate 'units' like a record deck, amplifier, cassette deck, speakers, etc. We can think of these as **building blocks** in our system. In fact we could draw a **block diagram** of the hi-fi system like this (Figure 1.1):

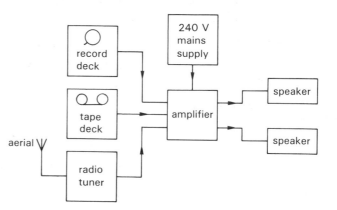

Figure 1.1 Block diagram of a typical hi-fi system

When we draw a block diagram we do not have to worry about how each block **works**. Usually we do not even worry about the exact **circuit** inside the block. All that is important is how each block **behaves** and how it **fits in** with the rest of the system.

Other systems

There are many other systems. Some are more complicated, like the national telephone system. Others are simpler, like a burglar alarm system. Every system can be understood, however, if you break it down into simpler blocks and see how they work together.

Main parts of a system (*Figure 1.2*)

- The **input device** sends a **signal** into the system.
- **Process circuits** take the **input signal** and perhaps change it or use it to control other circuits so that eventually an **output signal** is produced. Often the **process** part of a system can be broken down into several simpler circuits.

Modern ships rely heavily on electronic systems. This is the navigation system on the bridge of an oil rig supply ship.

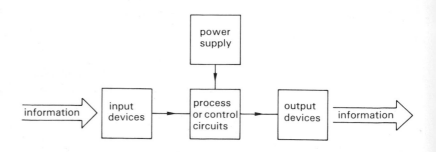

Figure 1.2 Block diagram of a typical system

Figure 1.3 The building blocks of systems are made from components. This is a 100 W amplifier. Notice the pins at the bottom to connect it to the rest of the system.

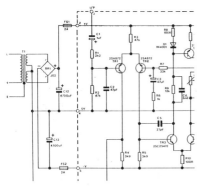

Figure 1.4 Part of the circuit diagram for the amplifier shown above

● The **output device** turns the output signal into something useful (like sound or a picture).
● The **power supply** provides the electrical energy to do all this.

Circuits

Of course an amplifier, for example, is not simply a box; it contains electronic circuits made up from electronic **components**. A circuit from an amplifier is shown in Figure 1.3.

Circuit diagrams

A **circuit diagram** is a drawing of how the circuit is made up, using **symbols** for the components. Figure 1.4 shows part of the circuit diagram for the circuit shown in Figure 1.3.

If you want to understand electronics you must learn something about how these components behave and their uses. There are only about a dozen or so basic kinds of components.

Once again, we do not have to know **how** each component works. Most of the time we only need bother about what each component can do and how it fits in with the rest of the circuit.

Summary

● An electronic system is made up from circuits. Each circuit is like a 'building block' of the whole system.
● A system is designed to do a specific thing. It has an input, process or control circuits, an output and needs a power supply.
● Circuits are built up from components.
● Each component must be in the right place if the circuit is to work properly. A circuit diagram shows how the components fit together.

Questions

1. What are the main parts of an electronic system?
2. What are electronic circuits made from?
3. Think of four examples of an electronic system. Try and draw a block diagram for each system.
4. In a TV set, what is the input, what is the output and where does the power come from?
5. For each of the following aspects of our lives, give an example of an electronic system that has helped to improve things:
   (a) road safety
   (b) house security
   (c) music
   (d) education
   (e) sport
   (f) help for people with disabilities

3

2
Practical Points

The best way to understand electronics is to build some circuits to see for yourself how they behave. Then you should try to combine simple circuits to make a system.

Figure 2.1 is a circuit diagram of a common circuit you will meet later in this book. There are many ways of actually building this circuit. Some methods are useful when first trying the circuit out. Other methods are usually only used when you want to make a permanent version.

'Push-in' boards

These are sometimes called **breadboards**. Components are pushed into holes on the board. (See Figure 2.2.) They are very useful for experimenting when trying out different components in circuits.

Using breadboards

The main thing to remember is that most of the underneath connections run **down** the board in groups of five. Figure 2.3 shows one way of building the circuit in Figure 2.1.

Blob boards

These are fibreglass boards covered with a pattern of copper pads. Components are connected by soldering them to the copper. (See Figure 2.4.) If you do not cut the leads, components can be removed and used again. Blob boards are useful for trying out circuits.

You can make your own blob boards by the 'printed circuit board' method (see below).

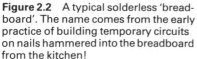

Figure 2.1 A circuit diagram of a simple transistor circuit

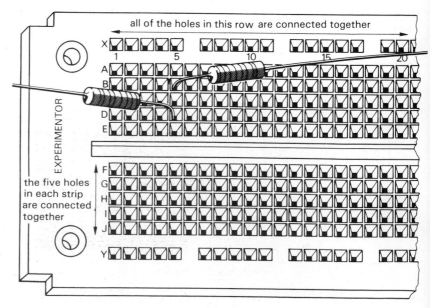

Figure 2.2 A typical solderless 'breadboard'. The name comes from the early practice of building temporary circuits on nails hammered into the breadboard from the kitchen!

4

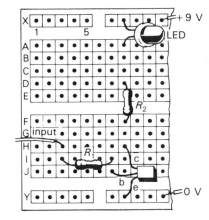

Figure 2.3 The circuit in Figure 2.1 built on bread-board

Space-saving techniques are used extensively in miniature equipment. In this personal stereo, small, leadless components have been soldered directly to the board.

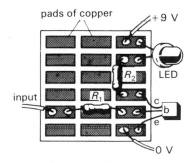

Figure 2.4 The circuit in Figure 2.1 built on blob board

Figure 2.5 The circuit in Figure 2.1 built on strip board (a) showing components (b) showing copper strips and soldered connections (c) showing how the components are mounted

Stripboard

This also has a pattern of holes and connections. Components are pushed through the holes and their leads are soldered to the copper strips on the other side of the board. (See Figure 2.5.) Stripboard is useful for making a permanent circuit when you have decided which components to use. One make of stripboard is called 'Veroboard'.

Using stripboard

Usually the strips do not go exactly where you want. Sometimes you may have to cut the strips, as shown in Figure 2.5 (b). Sometimes you may have to link two strips with a piece of wire.

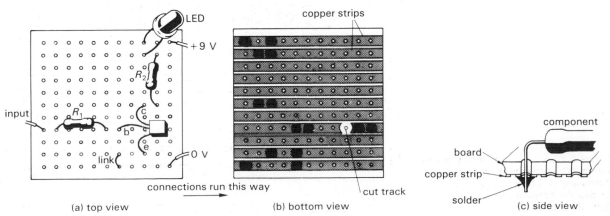

(a) top view   (b) bottom view   (c) side view

5

Printed circuit boards

A neater way of building circuits is to make a **printed circuit board** (**pcb**). You have to drill holes in the right places for the component leads. Copper connections run underneath the board. (See Figure 2.6.)

A typical example of a printed circuit board before the components are put on.

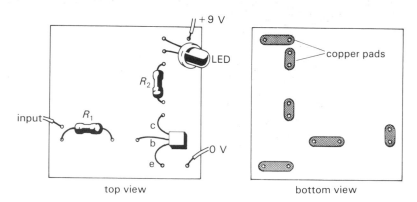

top view   bottom view

Figure 2.6 The circuit in Figure 2.1 built on a specially made printed circuit board. *Left* showing components; *right* showing copper pads.

Almost all systems you buy are built on printed circuit boards. A pcb is usually made for one particular circuit only. A different circuit will need a different pcb.

Making a PCB

1 A pcb starts off with a full covering of copper. The pattern of connections you want is printed on the copper with special ink which is not affected by acid.
2 The board is then put into acid or **ferric chloride**. This dissolves away the unwanted copper.
3 The board is then washed and the ink cleaned off leaving the copper only where you want it.
4 After the board has been checked, the holes are drilled.

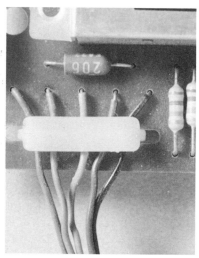

Connecting circuits together

Circuits can be joined together by soldering wires between them. Another way is to wrap the wires tightly around special pins using a special tool. This makes a good connection without soldering. (See Figure 2.7.)

Sometimes manufacturers bring all the connections to the copper pads along one edge of the board. The board is then plugged into a special **edge connector** to join it into the rest of the system. This makes it easy to change a board if it goes wrong. (See Figure 2.8.)

Figure 2.7 (*top*) Connecting wires soldered to the board (*bottom*) Wires connected by wrapping round special pins

Figure 2.8 The edge of this board has been designed to plug into a socket. Edge connectors like this are often used in home computers.

Summary

- Breadboards and blob boards are useful for experimenting with circuits. The components can be used over and over again.
- Stripboards are a quick way of making a permanent circuit.
- Printed circuit boards have to to be specially made for a particular circuit. They are neat and reliable.

7

3
Voltage and Circuits

There are several terms used in electricity and electronics which sometimes cause confusion.

Electromotive force (e.m.f.)

A battery can provide an electromotive force or e.m.f. (Figure 3.1). This e.m.f. can drive an electric current round a circuit connected to the two ends of the battery. E.m.f. is measured in volts (V). Other sources of e.m.f. are the mains, dynamos and solar cells.

Potential difference (p.d.)

Water can flow between two places if there is a difference in height between them. Electric current will only flow between two points in a circuit if there is a **potential difference** (**p.d.**) between them. Look at Figure 3.2. A p.d. exists here because a battery is providing an e.m.f. Potential difference is measured in volts.

Figure 3.1   This battery has an e.m.f. of 9 V

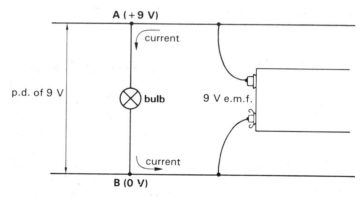

Figure 3.2   The e.m.f. of the battery produces a p.d. across the bulb

Voltage

Look again at Figure 3.2. We can choose some point in the circuit (usually the negative side of the e.m.f.) and call it 'zero volts' (0 V). Then, to find the **voltage at A** we measure the **p.d. between A and the 0 V point.**

If this sounds complicated remember that when a pilot says 'we are flying at a height of 5000 ft' what is really meant is that the distance between the plane and the sea is 5000 ft.

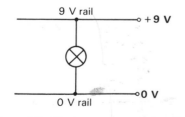

Figure 3.3   The two power supply rails

Rails

Since any point along the top wire in Figure 3.2 will be at a voltage of +9 V, we will call this wire the **+9 V rail**. Similarly the bottom wire is called the **0 V rail**. To keep diagrams simple, often we shall not show the source of the e.m.f. but just label the two power supply rails, as in Figure 3.3.

Voltmeter

This is an instrument for measuring p.d. It is connected across the p.d. we want to measure. See Figure 3.4.

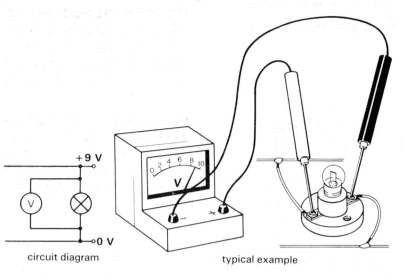

Figure 3.4 Measuring the p.d. across a bulb with a voltmeter

circuit diagram

typical example

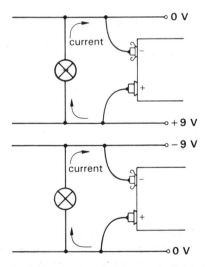

Figure 3.5 Two ways of showing negative p.d.

Negative voltages

If we reverse the connections to the battery we shall have to change the labels on the rails. There are two ways of doing this (see Figure 3.5). In each case the top rail is 9V **less** than the bottom rail. A current flows just as before, but in the other direction.

Voltages round a simple circuit

Look at Figure 3.6. In both of these circuits the voltage at A (V_A) is +9V and the voltage at B (V_B) is 0V. What is the voltage at C (V_C)?

Answer: In circuit (a), V_C is +9V. In circuit (b), V_C is 0V.

Explanation: In circuit (a), since no current flows through the bulb, there can be no p.d. across the bulb. Therefore if $V_A = +9V$ then $V_C = +9V$ also.

Current flows in circuit (b) because there is a p.d. across the bulb. Point C is now connected to the 0V rail, so $V_C = 0V$.

You might have thought that in circuit (a), because there is a bulb between A and C, V_C might be slightly less than V_A. This is **not** the case. When no current flows the voltage at each end of the bulb is exactly the same.

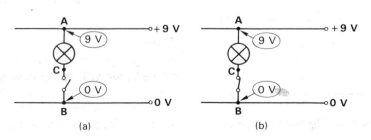

Figure 3.6 Voltages round a simple circuit with (a) the switch open (b) the switch closed

9

Summary

- An e.m.f. drives a current round a circuit.
- We speak of the voltage **at** a point.
- We speak of the p.d. **between** two points or **across** a device.
- If there is no p.d. across a device then no current flows through it.

Questions

1 What units is e.m.f. measured in?
2 What things can provide an e.m.f?
3 What is a voltmeter used for?
4 What will the voltmeter read in each of the following cases (Figure 3.7)?

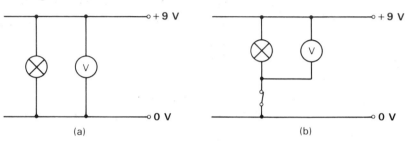

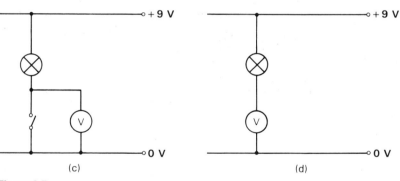

Figure 3.7

5 Which of the bulbs will light up in the circuit shown in Figure 3.8?

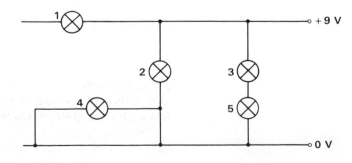

Figure 3.8

4
Current and Resistance

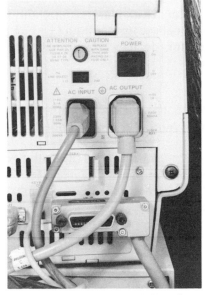

Mains connectors on a VDU. Apart from the letters 'AC', how else can you tell that this equipment needs an alternating current supply?

Current

In general a **large e.m.f.** will drive a **large current** round a circuit.

The amount of electric **current** flowing in a circuit is measured in **amps** (short for ampere) or **A** for short.

Milliamps

One amp is a rather large current and in most electronic circuits the current is much less. One **milliamp** (1 mA) is one thousandth of an amp, so

$$1000\,\text{mA} = 1\,\text{A} \text{ (or } 1\,\text{mA} = 0.001\,\text{A)}$$

For even smaller currents, units of **microamps** (μA) are used:

$$1000\,\mu\text{A} = \text{mA}$$

A.c. and d.c.

The current driven by the e.m.f. of a battery always flows the same way through the wires (from $+9\,$V to $0\,$V). This is direct current (or d.c.).

The current driven by the mains **alternates**. That is it flows first one way then the other. This is because the e.m.f. of the mains alternates. See Figure 4.1. This type of current is called alternating current (or a.c.).

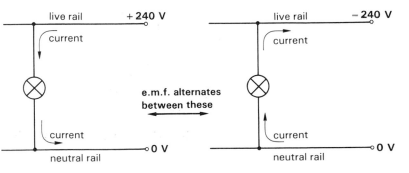

Figure 4.1   The alternating e.m.f. of the mains drives an alternating current (a.c.)

Frequency

When the current has changed direction and then changed back to where it was again, it has gone through one **cycle**.

The **frequency** of an alternating current is the number of times it goes through a complete cycle every second. The frequency of the mains in Britain is 50 cycles per second, or 50 hertz (50 Hz).

Huge ceramic insulators are needed to stop the current in overhead power cables from flowing where it should not.

Conductors and insulators

Some materials (such as metals) will conduct electricity easily. Even a tiny p.d. will make some current flow through them. These are called conductors.

Other materials (such as glass) will not conduct unless you put a very large p.d. across them. These are called insulators.

Semiconductors

Some materials, in particular **silicon and germanium**, are rather special. They are **semiconductors**. Devices made from these materials can behave like conductors or like insulators, depending on what we do to them. We can **control** how they conduct.

Some examples of **semiconductor devices** are transistors, diodes and of course **silicon chips**.

Resistance

When a p.d. is connected across a conductor, current flows through the conductor. The exact amount of current depends on the **resistance** of the conductor.

The resistance of a conductor is a measure of how difficult it is for an electric current to flow through it. A short, thick piece of wire has very little resistance. A long, thin wire, like a filament of a bulb, has more resistance. (See Figure 4.2.)

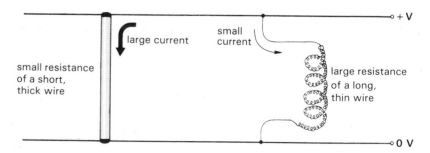

Figure 4.2   The current flowing depends on the resistance of the conductor

new symbol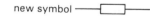

old symbol

Figure 4.3   Resistors: new and old circuit symbols and example of a typical device

Units of resistance

Resistance is measured in **ohms** (or Ω for short).

A conductor with a resistance of 1 ohm needs 1 volt to drive a current of 1 amp through it.

1 ohm is rather a small resistance. Most resistances we shall meet are much larger:

1 kilohm (1 kΩ) = 1000 Ω
1 megohm (1 MΩ) = 1000 kΩ

Resistors

A resistor (see Figure 4.3) is a component which is specially made to have a certain resistance.

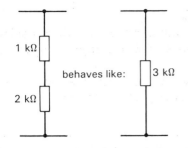

Figure 4.4   Two resistors in series

Resistors in series

If we connect two resistors in series the current has to go through both of them. Look at Figure 4.4. The total resistance here is 3 kΩ.

Potential dividers

Look at the circuits in Figure 4.5. Circuits like these are called **potential dividers**. The resistors are dividing the p.d. between themselves.

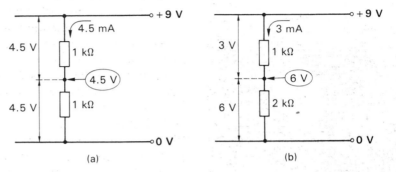

Figure 4.5   Two potential dividers

Note:

● Although both circuits had the same total p.d., this has been shared out between the two resistors in different ways. The largest resistor has the most p.d. across it.
● In both circuits, the **current flowing** can be calculated by **dividing the p.d. by the total resistance**. You can find out more about calculations like this in chapter 80.

Summary

● Current is measured in amps (A).
● The amount of current flowing between a p.d. depends on the size of the p.d. and the resistance of the device between it.
● Metals (like wires) have much less resistance than insulators (like plastic).
● Resistance is measured in ohms (Ω).
● A potential divider is made by connecting two resistors in series.
● The largest resistor in a potential divider has the most p.d. across it.

Questions

1   What units are each of the following measured in?
    (a) current   (b) p.d.   (c) resistance.
2   What is a semiconductor?
3   Explain these words: 'An **a.c.** with a **frequency** of **50 Hz**'.
4   Why did less current flow in the circuit in Figure 4.5(b) than that in Figure 4.5(a)?
5   Which of the circuits in Figure 4.6 has the largest voltage at point A?

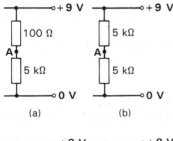

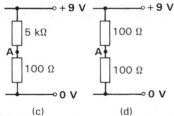

Figure 4.6

5
Effects of Current

Heat

When current passes through a resistance, electrical energy is converted into heat energy. More current produces more heat. Sometimes this is useful, but other times it is a nuisance or even dangerous unless we take precautions.

We cannot avoid this heat but we can make sure that it is kept under control. When choosing which resistor to use in a circuit, we must be sure that the resistor can stand the heat likely to be produced. We must be equally careful with other components which have resistance and carry current, e.g. transistors. (See chapter 74.)

Magnetic effect

Whenever electric current flows through a wire a **magnetic field** is produced. This field can be increased by winding the wire into a coil round a piece of iron, making an electromagnet.

Sometimes this is used to attract another piece of iron, as in electromagnetic relays. In other devices, such as meters and loudspeakers, this field is used to react against another magnetic field (often from a permanent magnet) and so produce movement (Figure 5.1).

The effects of electric current are used in different ways. How many can you see in this room?

Without vents to let the heat escape, the heating effect of electric current would quickly damage this TV set.

One of the dangers of electricity. This room was destroyed by a fire caused by an electrical fault.

A microphone uses electromagnetic induction to turn sound into electrical signals. Here a naturalist is recording the songs of birds. Why do you think the microphone has a large dish behind it?

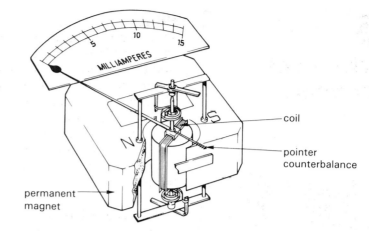

Figure 5.1   A meter uses a coil and magnet working together to give movement

coil

pointer
counterbalance

permanent
magnet

Electromagnetic induction

There is also an important opposite effect. A **changing** magnetic field can produce an electric current.

When a coil is in a steady magnetic field nothing happens, but if this field changes, an e.m.f. is produced across the ends of the coil. This effect is called **electromagnetic induction** and is used in many devices.

Chemical

When electric current passes through certain liquids it can cause a chemical change. This is the basis of electroplating and is discussed in detail in most chemistry and physics textbooks.

A record pick-up uses electromagnetic induction to turn vibrations into electrical signals.

The bodyshell of this car is electro-coated with a primer before being painted. This 'forces' the primer on to all surfaces, particularly in hidden and otherwise inaccessible areas. The electrical connection can be seen at the front.

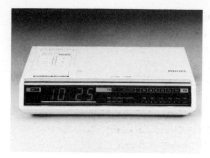

Light-emitting diodes (LEDs) use very little current and do not wear out like bulbs. This makes them ideal for digital displays like the one on this clock.

Light

In 1962 Nick Holonyak discovered that when electric current flows between the junction of two special materials (gallium arsenide and phosphorus) some light was produced. This light is not very bright but it can be useful. A light-emitting diode (LED) uses this effect.

Biological

Electric current can make certain living cells contract. This is used by animals to transmit **nervous impulses** and is described in many biology textbooks. It is also one of the effects of an **electric shock** (the other effect being burns caused by the heating effect). For an average adult, a current of about 40 mA can be fatal.

Power

We have seen that an electrical device converts electrical energy into other kinds of energy. The amount of energy that it converts each second is called the **power** of the device. Power is measured in **watts** (or W) and is calculated by multiplying the p.d. across the device by the current flowing through it:

power = p.d. × current

or   watts = volts × amps

Summary

● When current flows through a resistance, heat is produced.
● This heat can be useful, but mostly it is a nuisance.
● When current flows through a coil, a magnetic field is produced. This is used in speakers and relays.
● When the magnetic field is changed near a coil, an e.m.f. is produced. This is used in microphones and tape heads.
● The power of a device describes how much energy it changes each second. watts = volts × amps.

Questions

1   Name three effects of an electric current.
2   How can we make sure that the heat produced in a system does not cause damage?
3   Which devices use the magnetic effect of a current?
4   Which devices produce an e.m.f. by electromagnetic induction?
5   Copy out this table and fill in the missing values:

device	p.d. applied	current flowing	power
buzzer	12 V	0.5 A	
computer		3 A	15 W
amplifier	40 V		20 W
2.2 kΩ resistor	9 V		

6
Exercises

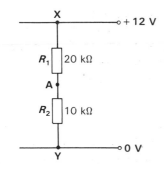

Figure 6.1

1   Here is a list of materials: **iron, silicon, PVC, glass, copper, silver, germanium, fibreglass, tin, carbon, lead.**
(a) Put this list of materials into three groups: 'conductors', 'insulators' and 'semiconductors'.
(b) Which of these materials are used in a printed circuit board?
(c) Which of these materials are used in transistors?
(d) Which of these materials are used in an electromagnet?
(e) Which of these materials are used in solder?

2   What units are the following measured in:
(a) e.m.f.
(b) current
(c) power
(d) resistance
(e) p.d.
(f) voltage?

3   Copy the circuit diagram shown in Figure 6.1.
(a) Label the 'positive rail' and the '0 V rail'.
(b) What do we call the arrangement of the resistors R_1 and R_2?
(c) How big is the p.d. between points X and Y?
(d) How big is the total resistance?
(e) How much current flows through the circuit?
(f) Which resistor has the largest p.d. across it?
(g) How big is the voltage at point A?
(h) If R_2 was replaced by a 100 kΩ resistor, what would happen to the amount of current flowing?
(i) What would happen to the voltage at point A?

4   Here are some circuits and devices (Figure 6.2):

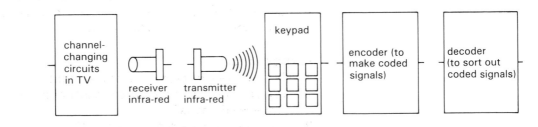

Figure 6.2

Draw a block diagram to show how they might be used to make an infra-red remote control system for a TV.

5   Put each of these stages in making a printed circuit board into the correct order:
(1) Put the board into ferric chloride solution.
(2) Check the board to make sure that you have not made any mistakes.
(3) Clean all the grease off the board.
(4) Clean off all the etch-resist and wash and dry the board.
(5) Draw the circuit on the board with etch-resisting ink.

17

6 Figure 6.3(a) is an example of a simple circuit built on a printed circuit board. Look at it carefully and compare it with the four layouts shown in Figures 6.3(b), (c), (d) and (e):

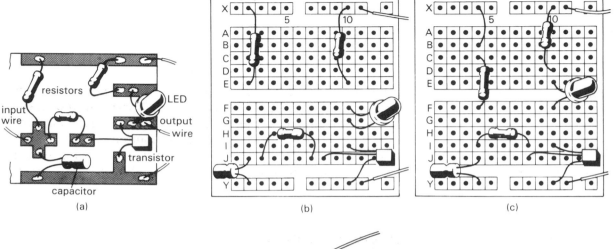

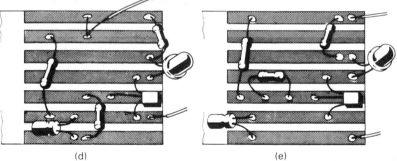

Figure 6.3 A circuit on (a) circuit board; (b) and (c) breadboard; (d) and (e) stripboard

(a) Which layouts are the same circuit as shown in Figure (a)?
(b) Where should the input and output wires be connected?

Section 2

Switching Systems

In chapter 3 current was controlled by a switch. Switches are widely used but they are not always ideal. They need someone (or something) to operate them and they eventually wear out. There are many systems in which current is controlled automatically. Street lights, for example, come on when it gets dark; nobody turns a switch so how is it done? The answer is to use an **electronic switch**.

Electronic switches

An electronic switch is one of the simplest **building blocks** in a system. Unlike a mechanical switch (Figure 7.1) which has metal contacts which are pushed together when the switch is operated, electronic switches have no moving parts. The simplest type has two connections called **input** (or **control**) and **output**. Being electronic it also needs two connections to the power supply rails. See Figure 7.2.

Street lights are controlled by electronic switches. The light sensor on the top of the lamp sends a signal to the switch when darkness falls.

Figure 7.1 Action of a mechanical switch

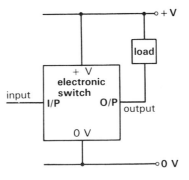

Figure 7.2 An electronic switch controlling a 'load'. The load could be a bulb, motor etc. or even another circuit.

State

When describing whether an electronic switch is **on** or **off** we often use the word **state**, e.g. 'the switch is in the off state'.

19

There are many different types of sensor. This one responds to infra-red rays given off by people. What do you think it is used for?

Using an electronic switch

An electronic switch is controlled by a signal which is sent into the input (Figure 7.3):

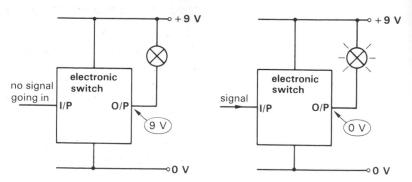

Figure 7.3 Action of an electronic switch

How an electronic switch behaves

When the switch is in the **off state** the voltage at the output is the same as the positive rail. This means there is no p.d. across the load and no current can flow through it.

When a signal is sent into the input the electronic switch goes into the **on state**. The voltage at the output drops to 0V. This produces a p.d. across the load and current can now flow through.

Input signal

The signal used to control the switch (i.e. the **input signal**) is usually a small electric current. This often comes from some kind of **sensor**. There are many types of sensor which can be combined with an electronic switch to make useful systems.

Light sensor

A **light-dependent resistor** (LDR) can be used to detect changes in light. (See Figure 7.4.) An LDR has a **large** resistance in the **dark** but it has a **small** resistance when light shines on it.

An example of a system which uses this is at the checkout of some supermarkets (Figure 7.5):

Figure 7.4 Light-dependent resistor (LDR): symbol and typical example

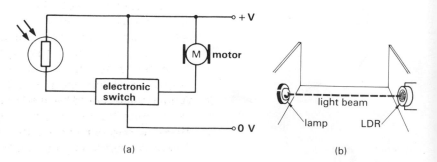

(a)   (b)

Figure 7.5 The conveyor belt at a supermarket checkout. (a) Block diagram of the system (b) Typical example of the arrangement used

A beam of light is sent across the conveyor belt to an LDR. If there is nothing in the way, the LDR has a small resistance. This lets an input current flow into the electronic switch and this turns the conveyor motor on. What do you think happens when a tin blocks the light beam and prevents it from falling on the LDR?

Heat sensor

A **thermistor** can be used as a heat sensor. (See Figure 7.6.) When it is **cold** it has a **large** resistance, but if we **heat** it up it has a **small** resistance. A fire alarm could use a thermistor (Figure 7.7).

When the thermistor gets hot, it has a small resistance. This lets current flow into the electronic switch which turns the alarm on.

Size of the current

The current needed as an input signal can be very small (say 1 mA) but the current flowing through the output can be much larger (say 1 A). So, a small, delicate device like an LDR can, indirectly, control a much larger device like an alarm.

Controlling larger currents

The simple electronic switch cannot stand really large currents. If we want to control very large devices such as motors or mains-powered equipment we can use a **relay**. In Figure 7.8, the electronic switch is used to turn the relay on by sending a small current through its coil. This makes the contacts move together and so completes the circuit for the motor.

You can read more about relays in chapter 79.

Summary

● Electronic switches use a small (input) current to control a larger current.
● The input current can come from a number of devices.
● Because they have no moving parts electronic switches do not wear out like mechanical switches.
● A relay can be added to control really large currents.

Questions

1 In which of these systems do we have to use an electronic switch rather than a mechanical switch:
   (a) turning on the lights in a room,
   (b) automatic street lights,
   (c) remote control of a TV,
   (d) turning on a radio by a knob,
   (e) a digital alarm clock turning on a radio in the morning,
   (f) a robot carrying out a programmed piece of welding?
2 Describe what happens to the voltage at the output of an electronic switch when the input receives an input signal.
3 What devices could be used to detect the following:
   (a) change in temperature   (b) change in light level?
4 Think of another use for the system shown in Figure 7.5(a) (try using something apart from a motor on the output).

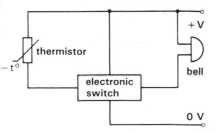

Figure 7.6   A typical thermistor

Figure 7.7   Block diagram of an automatic fire alarm

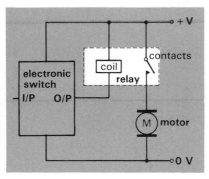

Figure 7.8

8
The Transistor

Transistors

The transistor is a component which can be used as a simple electronic switch. It has three leads (see Figure 8.1).

The **base** is used as the **input**. The current flowing into the base (perhaps from a sensor or even another circuit) will then control the current flowing at the **output** (the **collector**).

How a transistor behaves

If a transistor is connected between a bulb and the 0 V rail we can turn the bulb on and off by controlling the current flowing into the base of the transistor (Figure 8.2):

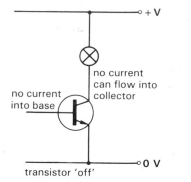 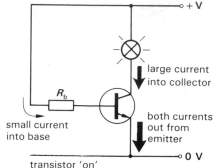

Figure 8.2   Current flow in a transistor.

Figure 8.1   Transistors: symbol and some typical examples. The larger ones can stand more current.

The resistor (sometimes called the **base resistor** or R_b) is used to make sure that the current flowing into the base is small. Leaving this resistor out would allow too much current to flow and the transistor may overheat and fail.

Voltages round a transistor

If we measure the voltage at the base and collector we will find:

	voltage at base	voltage at collector
transistor off	0 V	9 V
transistor on	0.7 V	0 V

When the transistor is **off**, the collector voltage is **high**. When the transistor is **on**, the collector voltage is **low**. Notice that we need less than 1 V at the base to turn the transistor on. In fact the base voltage will **never** be more than 0.7 V. The resistor R_b takes the rest of the power supply voltage.

A rain alarm

A simple **rain alarm** can be made by using a **water sensor** and a transistor. See Figure 8.3.

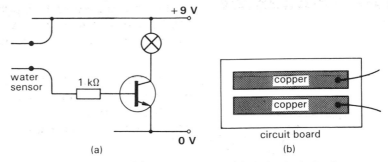

Figure 8.3 A rain alarm (a) Suitable circuit (b) A simple design for a water sensor

The water sensor can be a piece of circuit board with two tracks and a gap between them. In the dry no current can flow across the gap but if water bridges the gap it will let a small current flow. This current on its own would not be enough to light a bulb, but it is easily enough to control a transistor which can then control the bulb.

Summary

- A transistor has three leads: base, collector and emitter.
- The base can be used as an input, the collector can be used as an output.
- When a transistor is in the **off state**, its collector voltage is the same as the power supply.
- When a transistor is in the **on state**, its collector voltage goes down to 0 V.

Questions

1   When a transistor is used as a switch, which lead is used for the input and which lead is used for the output?
2   Copy this circuit diagram and fill in the gaps in the results table:

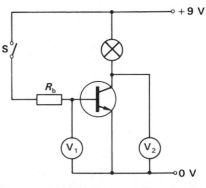

switch S	bulb	V_1 reads	V_2 reads
open			
closed	on		

results table

3   Draw a circuit diagram which uses an LDR to control the current flowing into the base of a transistor.

Electronics can help to make sure that plants receive just the right amount of water and light to grow successfully. In this photograph, a probe, which is connected to a meter, is inserted into the soil to measure water content.

23

9
The Transistor as a Switch

We will now look more closely at how a transistor switch can be used by trying to design an automatic light-sensing system.

Simple light-sensitive switch
The simplest circuit is shown in Figure 9.1. You will see that it is very similar to the rain alarm shown in chapter 8 but with an LDR in the place of the water sensor.

How it works
In the dark: the LDR has a large resistance and no current can flow into the base of the transistor, so it is in the **off** state.
In the light: the LDR has a small resistance so now some current can flow into the base of the transistor and it is in the **on** state.

This could be used to turn a radio on when it gets light in the morning—provided you do not mind getting up at dawn!

Limitations
In practice this circuit is not very good because:
● It needs a large change in light levels to operate.
● We cannot adjust the amount of light needed to turn it on.
Trying to control a transistor by controlling the base current is not generally satisfactory.

Voltage control of transistors
A better way is to control the **voltage** at the left-hand end of resistor R_b. If this voltage is more than about 1 V then enough current will flow into the base to turn the transistor on.

Note on high and low voltages
For the rest of this section on electronic switches we shall simply call a voltage **high** if it is **1 V or more** and **low** if it is **less than 1 V**.

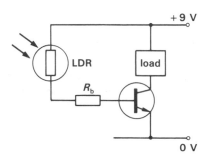

Figure 9.1 Circuit for a simple light-operated switch

Hundreds of lights are needed at the Mossmorran Natural Gas Plant shown here. It is hardly surprising that they are all controlled automatically!

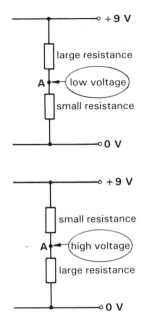

Figure 9.2   Two potential dividers

Potential divider

The potential divider in chapter 4 is a good way of controlling voltage. The voltage at A in Figure 9.2 depends on the values of the two resistors.

A better light-sensitive switch

In the circuit in Figure 9.3 the LDR and R_2 make a potential divider. Remember we are trying to make the voltage at A at least 1V to turn the transistor on.

How it works

In the dark the resistance of the LDR is much more than the resistance of R_2, so the voltage at A is nearly 0V and the transistor is in the off state. If we let more light fall on the LDR then its resistance will decrease. This makes the voltage at A rise. If enough light falls on the LDR, the voltage at A will be more than 1V and the transistor will switch into the on state.

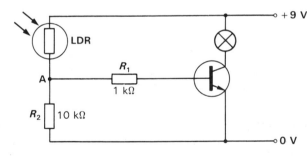

Figure 9.3   An improved light-sensitive switch

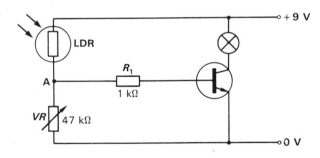

Figure 9.4   Light-sensitive switch with adjustment facility

Improvements

The resistor R_2 must be chosen carefully. If R_2 is too small then even if the LDR is in full sunlight, its resistance might never become small enough for the voltage at A to reach 1V.

The best way is to use a variable resistor for R_2. This lets you adjust the value of R_2 until the circuit operates at exactly the light level required. See Figure 9.4.

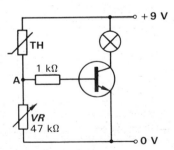

Figure 9.5   Circuit for a heat-sensitive switch

Heat-sensitive switch

The fire alarm described in chapter 7 could be made with the circuit shown in Figure 9.5.

When the thermistor is heated then its resistance gets less. The voltage at A rises. If it reaches 1V the transistor will turn on.

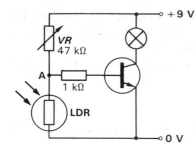

Figure 9.6 Circuit for a shadow-operated switch

A shadow detector

In Figure 9.6 we have exchanged the places of VR and the LDR. Now, allowing **less** light to fall on the LDR will make the voltage at A rise. If VR is adjusted correctly then the system will turn on when a shadow falls on the LDR.

By changing the positions of the components in the light sensor we have made the system operate in the opposite way. This is the basis of some burglar alarms, automatic street lights and so on.

Summary
- It is better to control a transistor switch by controlling the voltage at the input.
- A **high** voltage is **1 V or more**. A **low** voltage is **less than 1 V**.
- A potential divider can be used to provide this voltage.
- An LDR and a variable resistor can be used as a **light sensor** by connecting them as a potential divider.
- If the positions of the variable resistor and the LDR are exchanged, the light sensor becomes a **shadow sensor**.

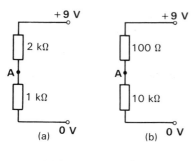

(a) (b)

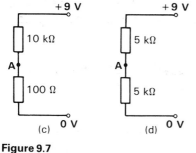

(c) (d)

Figure 9.7

Questions

1 Is it better to try to control the current flowing into the base or to control the voltage at the end of R_b?

2 Which of the potential dividers in Figure 9.7 will have a voltage at A bigger than 1 V?

3 Copy the circuit diagram in Figure 9.8 and answer the following questions:
   (a) How big is V_A when the LDR is in the dark?
   (b) How big is V_B when the LDR is in the dark?
   (c) Is the transistor on or off?
   (d) How big is V_C when the LDR is in the dark?
   (e) What happens to V_A, V_B and V_C when the light shines on the LDR?
   (f) How can the circuit be adjusted to react to less light?
   (g) How can the circuit be altered to turn the transistor on when the LDR is in the dark and turn it off in the light?

4 A particular automatic street lamp switches on when the voltage from its light sensor exceeds 1.0 V. During one evening, the output from the light sensor varied thus:

time of day	7.00	7.30	8.00	8.30	9.00	9.30
output from sensor	0.2 V	0.2 V	0.5 V	0.8 V	1.5 V	1.5 V

(a) Draw a graph of the output of the sensor against time.
(b) Estimate from your graph when the street lamp switched on.

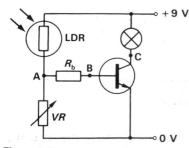

Figure 9.8

10
The Inverter

In the last chapter we changed a light detector system into the opposite (a shadow detector system) by changing round the components in the sensor part of the system.

Instead of doing this we could have kept the sensor as it was and used an **inverter**. Figure 10.1 shows the symbol for an inverter.

How an inverter behaves

A **low** voltage on the input makes a **high** voltage appear at the output.

A **high** voltage on the input makes a **low** voltage appear at the output.

In short the output is the opposite (or inverse) of the input. The easiest way of describing this is in a **truth table** (Figure 10.2).

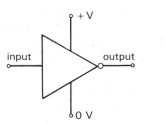

input	output
low	high
high	low

Figure 10.1   Symbol for an inverter. Often the power supply connections are not shown on block diagrams.

Figure 10.2   Truth table for an inverter

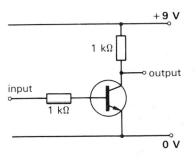

Figure 10.3   Circuit for an inverter using a transistor

Inverter circuits

The transistor switch can be used as an inverter (Figure 10.3). When the voltage at the input is **high** then the voltage at the output is **low** and vice versa.

Using an inverter in a shadow detector system

Here is a block diagram of the system (Figure 10.4):

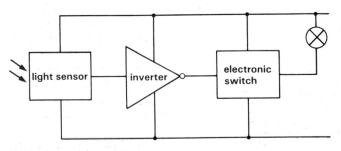

Figure 10.4   Block diagram for a shadow detector

When a shadow falls across the light sensor the input to the inverter goes **low**, so the input to the electronic switch goes **high**.

Circuit for a shadow detector

A suitable circuit for a shadow detector system is shown in Figure 10.5:

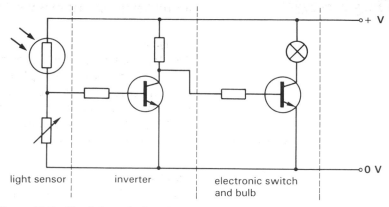

light sensor | inverter | electronic switch and bulb

Figure 10.5   Circuit for a shadow detector

Summary

● An inverter can be used to turn a 'high' signal into a 'low' signal and vice versa.
● A transistor switch behaves like an inverter.
● A truth table is a simple way of describing how a circuit behaves.

Questions

1   Draw the symbol for an inverter and its truth table.
2   Look at the circuit diagram in Figure 10.6 and then answer the following questions.

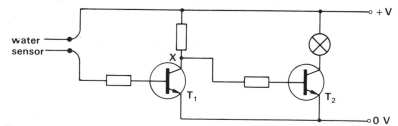

Figure 10.6

(a) When the water sensor is in the dry: (i) Will transistor T_1 be in the off state or in the on state?   (ii) Will the voltage at point X be high or low?   (iii) Will transistor T_2 be in the off or in the on state?   (iv) Will the lamp be on or off?
(b) Now explain what happens when the water sensor gets wet.
(c) Think of a use for this system.

3   The block diagram in Figure 10.7 is of a system which will operate a fan if the temperature rises. How would you change it so that it operated a heater if the temperature dropped? Draw the system and explain how it works.

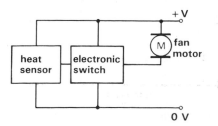

Figure 10.7

11
The Latch

An important point about all the circuits described so far is that they are only in the on state as long as the correct input is applied. If the input changes, the switch changes to the off state immediately. Sometimes we do not want this to happen.

Pulses

A **pulse** is a signal which goes suddenly from low to high and back again. We can draw a simple diagram to show this (Figure 11.1).

A pulse could be made by shining a light on a light sensor for a moment only. Other pulses can be made using push button switches or even special circuits called **pulse generators**. There are several circuits which are designed to be used with pulses.

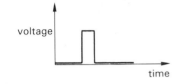

Figure 11.1 A voltage pulse

Latching

A **latch** will turn on and **stay on** once it has been triggered by a pulse.

How a latch behaves

The easiest way to describe the behaviour of a latch is to draw a diagram to show how the output voltage changes if we change the input voltage. This is known as a **timing diagram**. See Figure 11.2.

Silicon-controlled rectifier

A **silicon-controlled rectifier** (**SCR**) or **thyristor** (Figure 11.3) can be used as a latch. It will not conduct a current until a voltage (called a **trigger**) is put on the gate. In this respect it is like a transistor. Unlike a transistor, however, the SCR will continue to conduct even if the trigger voltage is removed from the gate.

Simple burglar alarm system

The circuit for a simple burglar alarm in Figure 11.4 shows how to use an SCR. It is triggered by an intruder breaking a beam of light.

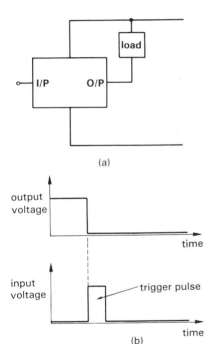

Figure 11.2 (a) Connecting a latch to a load (b) Timing diagram for a latch

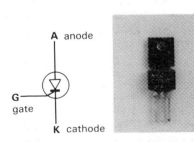

Figure 11.3 Silicon-controlled rectifier (SCR): symbol and typical device

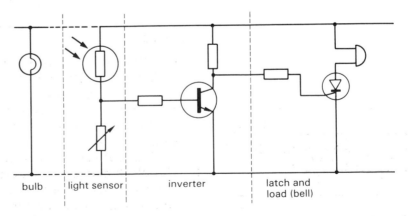

Figure 11.4 Circuit for a simple 'beam of light' burglar alarm

Resetting the SCR

Once triggered, the SCR can only be turned off or **reset** by:

● Turning off the power supply, **or**
● briefly connecting the anode to the 0 V rail.
● Connecting the anode to a negative voltage.

These methods generally need a mechanical switch. This makes the SCR ideal for systems like burglar alarms but if we need to reset the latch by an electronic signal a different type of latch is used. One example is a **bistable**. (See chapter 24.)

Summary

● A signal which is high only for a moment is called a **pulse**.
● A latch will turn on and stay on after a pulse at its input.
● An SCR can be used as a latch.

Questions

1   Draw a diagram to show what a pulse is.
2   Draw a diagram to show how a latch behaves when a pulse is sent into the input.
3   Which of the following systems need a latch to hold the output on after a pulse input?
    (a) a TV remote control
    (b) a temperature controller for a central heating system
    (c) an automatic street light
    (d) a system to turn a radio on when you clap your hands.
4   Draw a circuit of an SCR controlling a lamp and describe how the lamp can be turned on and off.
5   Would the circuit still latch if an a.c. supply were used rather than a d.c. supply? Explain your answer.
6   A typewriter or computer keyboard has a 'shift' key which must be held down while another key is pressed to produce capital letters. Disabled people could find this difficult or impossible. One solution could be to use a latch circuit as shown in Figure 11.5.
    (a) Draw a timing diagram to explain why they can now produce capital letters using only one finger.
    (b) Copy the system and add a key which will reset the latch.
    (c) Add an LED which will indicate when the shift key is pressed.
    (d) What other keys on computers or typewriters could usefully have this system connected to them?

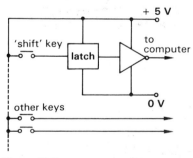

Figure 11.5

12 Transducers and Sensors

Transducers and **sensors** link (or **interface**) the system to the outside world. They are used at the input and output.

Input transducers and sensors

Anything that can be measured (pressure, light intensity, temperature, etc.) can be turned into an electrical signal by the right input transducer or sensor. Transducers produce their own e.m.f. Sensors use the e.m.f. from the system's power supply. The **signal** from both of them is usually a **change in voltage**.

Examples of input transducers and sensors are: reed switches and magnets, ultrasonic transmitters and receivers, microphones, infra-red sensors, vibration sensors and LDRs.

Output transducers

These turn electrical signals into something else. A speaker produces sound, a motor produces movement, etc.

Burglar alarm

A burglar alarm system needs many input devices. Some will be simple switches on doors and windows or pressure mats. Others might be transducers which detect body heat, movement or noise. Suppose you had to design burglar alarms for each of the places shown in these photographs. What sensors would you use?

The interior of an infra-red ultrasonic receiver-transmitter.

A luxury yacht.

A typical room in a house.

31

13
Designing a System

Now that we have seen a few 'building blocks' we can look at how to **design a system**. This is best done in stages:

1 Decide what the system is going to do.
2 Decide what the input and output transducers are going to be.
3 Look at the 'signal' we expect to get from the input.
4 Look at what 'signal' we want to have at the output.
5 Turn the input signal into the output signal which we want.
6 Draw a block diagram of the system.
7 Check to see that it will work as expected.

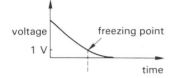

Figure 13.1 The output voltage from the heat sensor falls as the temperature drops

Example
Design a system to tell if there has been a frost in the night.

Design
1 A lamp will go on and stay on if the temperature goes below 0°C.
2 The input will be from a heat sensor. The lamp is the output.
3 We expect to get the signal in Figure 13.1 from a heat sensor as the temperature falls.
4 The lamp wants the signal shown in Figure 13.2.
5 (a) A latch will hold the lamp on once triggered.
   (b) The latch needs a high voltage for a trigger (Figure 13.3), so
   (c) we must invert the output from the heat sensor.
6 This system will do the job:

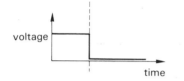

Figure 13.2 The signal needed by the lamp

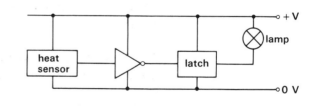

7 Check by drawing a timing diagram:

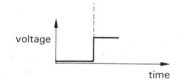

Figure 13.3 The signal needed at the input to the latch

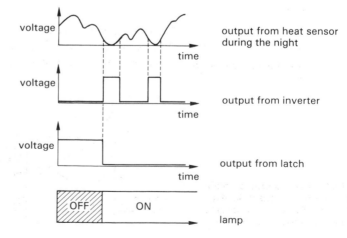

14

The Comparator

Sometimes we need to compare two voltages to find out if they are different. A **comparator** (Figure 14.1) can do this.

How a comparator behaves

A comparator has two inputs (labelled '+' and '−') and one output. If we put a voltage (call it V_1) on the '+' input and another voltage (V_2) on the '−' input, then the voltage produced at the **output** depends on how V_1 and V_2 **compare** with each other.

In Figure 14.2 we have fixed V_1 at 3 V by using a potential divider. We can vary the size of V_2 by the variable resistor.

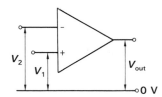

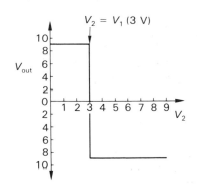

Figure 14.1 Symbol for a comparator, showing input and output voltages

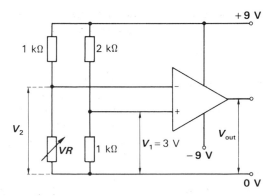

Figure 14.2 Circuit to investigate the behaviour of a comparator

The output will be −9V, 0V or +9V, depending on the size of V_2 (Figure 14.3).

Comparator circuits

The actual circuit inside a comparator is quite complicated but luckily this does not matter since comparators can be bought as **integrated circuits**. One common type is known as a '741'. (See Figure 14.4.) It contains about twenty transistors and resistors all inside a small case with eight leads. It costs about 20p.

Power supplies

Notice that to make it possible for the output voltage to be either +9V or −9V, two power supplies are needed.

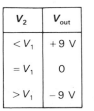

V_2	V_{out}
$< V_1$	+9 V
$= V_1$	0
$> V_1$	−9 V

Figure 14.3 Graph and table of results obtained

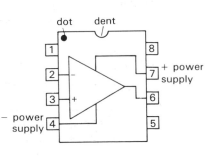

Figure 14.4 '741' integrated circuit: 'pinout' diagram and example of a typical device. The dent or dot in the case shows which is pin number 1.

Using a single power supply

If we do not need a negative output voltage then we can use a single power supply with the negative side connected to the 0 V rail as usual. The output will then only be 0 V or +9 V. (Figure 14.5) This is useful if we only want a change in output when V_2 actually becomes bigger than V_1.

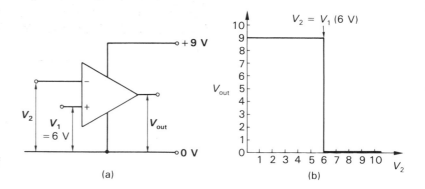

Figure 14.5 (a) Comparator working on a single power supply   **(b)** Graph of results obtained

(a)

(b)

Using comparators

A common way of using a comparator is to put a fixed voltage (called a **reference voltage**) on one input and connect the other input to a sensor.

The familiar light-operated switch can be made using a comparator like this (Figure 14.6):

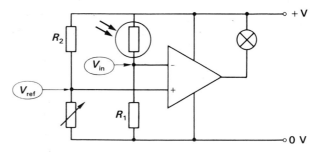

Figure 14.6 Circuit for a light-operated switch using a comparator

How it works

V_{ref} depends on the setting of the variable resistor. V_{in} depends on the amount of light falling on the LDR. In the dark, V_{in} is smaller than V_{ref}. The output is **high**. In the light, V_{in} is more than V_{ref}. The output is **low**.

Advantages of using a comparator

● It is easy to adjust the light level which will operate the switch. This is done simply by changing V_{ref} (by adjusting VR).
● It switches more quickly than the simple transistor switch.

Overload indicator

The circuit in Figure 14.7 can be found in some cassette decks. It helps to avoid distortion caused by trying to record a signal which is too large. It monitors the signal going in to the recorder. If this signal (V_{in}) exceeds V_{ref} then the comparator switches on a warning light, advising you to turn the recording level down! Notice that instead of a bulb, an LED is used as a warning light. LEDs take much less current than a bulb and are much smaller. (See chapter 75.) This circuit is in fact a simple **analogue-to-digital converter**.

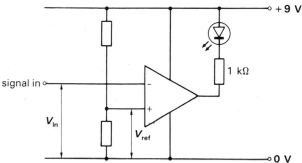

Figure 14.7   An overload indicator

Level indicator

An extension of the last circuit is shown in Figure 14.8. It is sometimes called a **level indicator**. It can give an indication of the size of a signal by turning on more LEDs as the signal increases. Each comparator, going up the chain, has a slightly higher V_{ref}, set by the 'ladder' of resistors.

Although this system may seem bulky and expensive, it can all be done using just two integrated circuits. One contains ten comparators, all sharing the same power supply. Another contains ten LEDs in a line. See Figure 14.9.

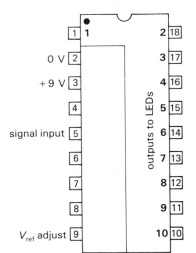
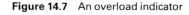

Figure 14.9   Pinout of an LM3914 bar/dot driver. The photograph shows a bar graph display using ten LEDs in one package.

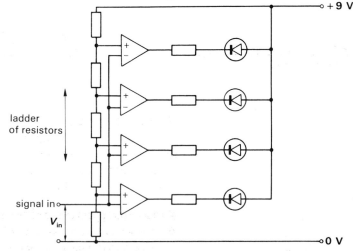

Figure 14.8   Level indicator using four comparators. Note that the power supply connections are not shown.

Summary
- A comparator compares two voltages and produces an output which depends on this comparison.
- The output can be a positive voltage, zero or a negative voltage. This means that two power supplies must be used.
- With one power supply the output can be positive or zero.
- Electronic switches using comparators are easy to adjust by changing the **reference voltage**.

Questions
1 Draw the symbol for a comparator and describe what it does.
2 Each of the signals in Figure 14.10 were connected in turn to the '−' input of the comparator shown. Copy each one and in each case draw another line on the graph to show what happens at the output. The first one has been done for you.
3 Look at the block diagram in Figure 14.11 and explain how it can tell you which light sensor is receiving most sunlight.

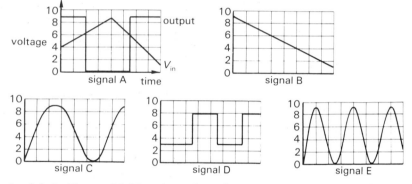

4 (a) In Figure 14.12, a thermistor is used in a potential divider to provide one of the inputs to a comparator. The other input is fixed at $+5V$. What is the smallest voltage needed at the non-inverting input to make the output go from low to high?
(b) Calculate the resistance the thermistor must have to produce this voltage.
(c) The thermistor resistance changes with temperature as shown in Figure 14.13. (i) If the temperature steadily falls, when will the LED go on? (ii) Calculate the value of R_1 needed if the LED is to go on at 20°C.

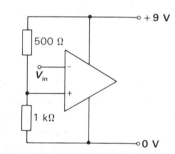

Figure 14.10

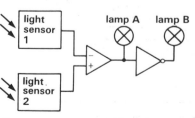

Figure 14.11

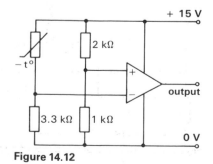

Figure 14.12

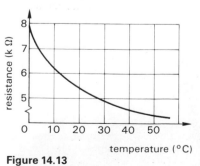

Figure 14.13

36

15
Time Delay Circuits

Zebra crossings can cause delays on busy streets. Pelican crossings with their timing circuits ensure that neither pedestrians nor the traffic have to wait for too long, and the warble sound tells blind people when it is safe to cross.

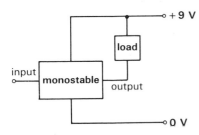

Figure 15.1 Monostable connected to a load

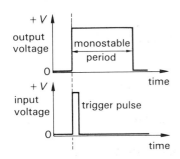

Figure 15.2 General shape of the timing diagram for a monostable

Sometimes we want a system to have some kind of time delay before the output changes. For example in a pelican crossing both the green man and the warble sound are on for about 15 seconds before they are turned off again. A circuit which can produce a time delay like this is called a **monostable** or **pulse generator** (Figure 15.1).

How a monostable behaves

Like other switch circuits, a monostable has two states (output high or low). The monostable, however, is only stable in **one** state (usually output low). When it receives a suitable input signal (or 'trigger') it is forced into the other state but **does not stay there**. It drops back to its stable state after a time delay (the **monostable period**). It produces a **voltage** or **trigger pulse**. A timing diagram is shown in Figure 15.2.

Example of using a monostable: a pelican crossing

Look at Figure 15.3. Normally the red lamp is on. When S is pressed, the green lamp and the warble sound go on for the monostable period (about 15 s).

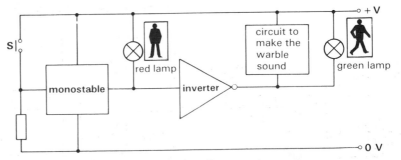

Figure 15.3 Block diagram of part of a pelican crossing system

Figure 15.4 Symbols for capacitors
(a) Normal type   (b) Electrolytic type

Capacitors

Monostable circuits use a component called a **capacitor**. The symbol for a capacitor is shown in Figure 15.4. It can be 'filled with electric charge' (or **charged up**) by the arrangement shown in Figure 15.5. As it charges up, the p.d. across it rises. If the charging current flows through a resistor this process takes some time, depending on the size of the capacitor and the resistor.

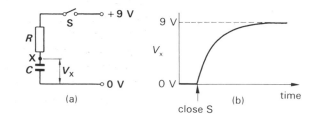

Figure 15.5 Charging a capacitor through a resistor   (a) Suitable circuit   (b) Graph showing change in the p.d. across the capacitor

The charged capacitor can also be discharged by connecting it across a resistor (see Figure 15.6). Once again the time taken depends on the size of the capacitor and resistor.

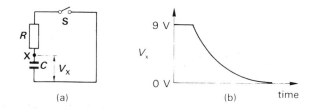

Figure 15.6 Discharging a capacitor (a) Suitable circuit   (b) Graph showing change in the p.d. across the capacitor

Simple time delay using one transistor

Figure 15.7 shows a simple time delay circuit. If S is opened there is a time delay before the lamp goes on.

When S is closed, the voltage at A (V_A) is 0V, the transistor is off and the capacitor is discharged. When S is opened the capacitor charges up through R and so V_A steadily rises. After a while V_A becomes more than 1V and so the transistor is turned on.

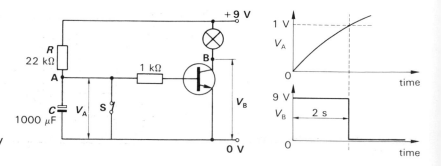

Figure 15.7 Simple monostable (delay before turning on) using a transistor

R (kΩ)	C (μF)	delay (s)
10	1000	1
47	1000	5
100	470	5
100	2000	20

Figure 15.8 Table of typical delay times for the monostable shown in Figure 15.7

Changing the delay time

With the component values shown in Figure 15.7, the delay time (or **monostable period**) is about 2 s. This can be increased by increasing the value of R or C. Figure 15.8 shows some typical values.

Another time delay circuit

In the circuit shown in Figure 15.9 there will be a time delay between opening S and the lamp being turned off. Once again the delay is caused by the capacitor taking time to charge up through R, only this time the voltage at A falls as the capacitor takes more and more of the 9 V p.d.

Eventually the capacitor will have more than 8 V across it, leaving less than 1 V at A and the transistor turns off.

Notice that the time delay is much longer than before even though the components are the same values.

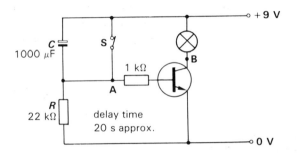

Figure 15.9 Simple monostable (delay before turning off) using a transistor

Using a comparator as a monostable

In the circuit in Figure 15.10 the lamp is normally off. If S is closed and then quickly opened again, the lamp goes on for about 5 seconds only. Once again this depends on the charging of the capacitor through R_2. The output switches when the voltage at the '+' input becomes more than the voltage at the '−' input. (Compare this to the 1 V needed in the transistor circuit.) This can give larger time delays.

The delay can be **adjusted** by making R_3 (or R_1) **variable** and so changing the **target** voltage at the '−' input.

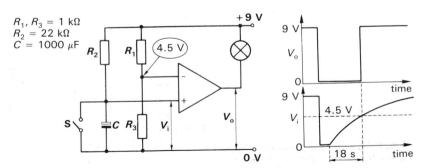

Figure 15.10 Monostable using a comparator

Summary
● A monostable is stable in one state only.
● It can be forced into its unstable state by a suitable trigger but returns to its stable state after a time delay.
● Monostables use the charging (or discharging) of a capacitor through a resistor to create the time delay.

Questions
1 How many states does a monostable have?
2 What is meant by the 'monostable period'?
3 Draw a timing diagram to show how a monostable behaves.
4 Copy the block diagram in Figure 15.11 and describe what happens when S is closed briefly and then opened again.

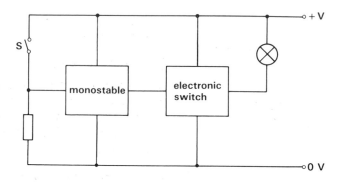

Figure 15.11

5 How could you alter the system to make it turn a lamp off when S is closed briefly? Draw a block diagram of the new system.

16
An Integrated Circuit Timer

There are several integrated circuits which are specially designed to be used as monostables. One type is the NE555, usually known as the 555 timer IC. (See Figure 16.1.) It can be triggered by a pulse which goes from high to low for a short while. The circuit in Figure 16.2 shows how it is connected up.

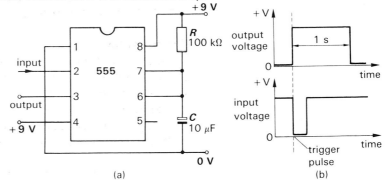

Figure 16.2 Monostable using a 555 timer IC   (a) Circuit   (b) Timing diagram

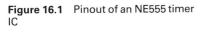

Figure 16.1 Pinout of an NE555 timer IC

R (kΩ)	C (μF)	T (s)
100	10	1
500	10	5
500	50	25
500	100	50

Figure 16.3 Table of approximate monostable periods (T) with various values of R and C

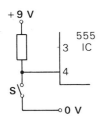

Figure 16.4 Close S to interrupt the timing process and reset the circuit

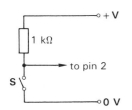
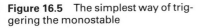

Figure 16.5 The simplest way of triggering the monostable

How it behaves

The stable state is with the output (pin 3) **low**. The timing diagram shows what happens when a trigger pulse goes into the input (pin 2). With the component values shown the monostable period is about 1 second. This can be increased by increasing R or C. (See Figure 16.3.)

Resetting the timer

The monostable period ends when the voltage at pin 7 (which has been rising as C charges up) reaches 6V. The circuit then automatically resets itself ready for the next input trigger pulse. We can also reset the circuit **before** the end of the monostable period by taking pin 4 down to 0V. (See Figure 16.4.)

Trigger pulse

The circuit needs a trigger pulse which goes from high to low. One way of doing this is to use a simple switch (Figure 16.5). We must release the switch before the monostable period ends or the circuit will not reset. We can also trigger the circuit by using **digital signals** (see section 3).

Questions

1   The monostable period (T) of the 555 timer circuit shown in Figure 16.1 can be found by:
$$T = 1.1 \times R \times C$$
(T is in seconds, R is in ohms and C is in farads.) Use this to calculate the monostable period when:
(a) $R = 1000\,\Omega$, $C = 470\,\mu$F   (b) $R = 33\,$kΩ, $C = 22\,\mu$F.

2   A burglar alarm is to ring for 5 minutes and then switch off. It is to be controlled by a 555 monostable. You have a 4700 μF capacitor. What value of resistor would you use?

41

17
The Astable

An **astable** (or **oscillator**) also has two states but it is **not stable in either state**. As long as power is supplied it constantly switches (or oscillates) from one state to the other. It does not have an input.

Figure 17.1 shows an astable and its timing diagram. The load (a lamp in this case) will be constantly switched on and off.

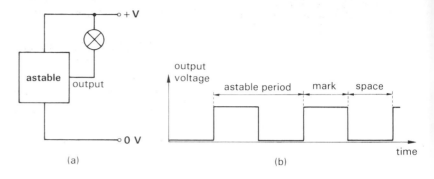

Figure 17.1 The astable (a) Connected to flash a lamp (b) Typical output signal from an astable

(a)   (b)

Flashing lights are easier to spot than steady lights, especially in difficult conditions. How do you think these warning lights are controlled?

Useful words

The **astable period** is the time it takes to change from one state to the other and back again (i.e., go through one **cycle**).

The **frequency** is the number of complete cycles it can go through in one second.

Uses of astables

● Flash a light on and off.
● Vibrate a loudspeaker and so produce sound.
● Provide a regular series of pulses (or **clock pulses**). These are used in systems like computers where a sequence of events must be strictly regulated.

Astable circuits

There are several different ways of making an astable. One of the easiest is to use a 555 timer IC again (Figure 17.2):

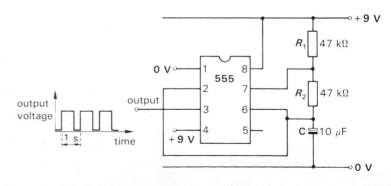

Figure 17.2 Connecting the 555 timer IC as an astable

How it works

This is similar to the monostable circuit (chapter 16) but the trigger input (pin 2) is connected to pin 6 so that the circuit keeps triggering itself, making it an astable. Each time it changes state it produces a click in the loudspeaker. If the frequency is fairly low (say 1 Hz) each click can be heard, but if the frequency is high (say 50 Hz or more) we hear a musical note.

Changing the frequency

The frequency of the astable depends on the values of the resistors and C. With the values shown the frequency is about 1 Hz. The frequency can be increased by decreasing the value of C or R_1 and R_2. Some typical values are shown in Figure 17.3.

Mark-to-space ratio

Figure 17.2 shows that the output is high for twice as long as it is low. We say that the **mark-to-space ratio** is **2:1**. For the 555 timer, this happens if R_1 and R_2 are the same value. We can increase the mark-to-space ratio by increasing the value of R_1 (Figure 17.4):

R_1 and R_2 (kΩ)	C (μF)	frequency (Hz)
47	10	1
10	5	10
10	0.5	100
1	0.1	2000

Figure 17.3 Table of typical frequencies obtained with various values of R and C

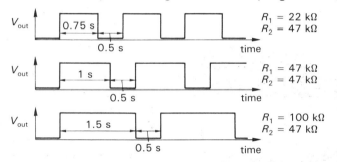

Figure 17.4 Timing diagrams to show how the mark-to-space ratio varies when R_1 is increased

Connecting astables in systems

The astable has no input of course. One way of controlling it is to switch its power supply on or off. We can connect it as we would connect any other load, for example:

Astable controlled by a monostable

An astable can be controlled by a monostable to produce a single burst of sound (Figure 17.5):

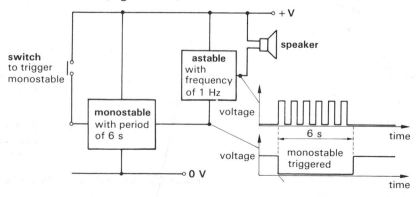

Figure 17.5 Astable controlled by a monostable to produce a single burst of sound

Controlling one astable with another *(Figure 17.6)*

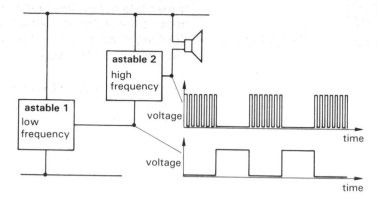

Figure 17.6 Two astables linked together to produce pulses of sound

The first astable pulses the second astable on and off. Interesting sound effects can be produced with this system depending on the frequencies of the two astables (f_1 and f_2):

$f_1 = 0.2$ Hz and $f_2 = 50$ Hz will give a sound like a **foghorn**,
$f_1 = 5$ Hz and $f_2 = 500$ Hz will sound like a **pelican crossing**,
$f_1 = 10$ Hz and $f_2 = 2$ kHz will sound like a **chirping bird**.

Other astable circuits

The 555 timer is not the only way of making an astable circuit. Transistors can be used, for example (see next chapter). The astable is one of the most useful building blocks of all.

Summary

- Astables keep switching from one state to another.
- They can be used to flash lights, make sound or provide pulses to control other systems.
- A 555 timer IC can be used as an astable.
- The frequency can be increased by decreasing the values of the timing resistors or capacitors.

Questions

1 Draw a block diagram showing an astable being used to flash a lamp on and off.
2 Draw a diagram of the output signal from an astable and use it to explain the meaning of the following:
   (a) astable period   (b) frequency   (c) mark-to-space ratio.
3 Draw a block diagram of a system which will produce a 1 kHz note from a speaker for 10 seconds when a beam of light is broken. Draw timing diagrams to explain what happens at the output of each block of your system.

18
Multivibrators

Multivibrators are an important group of circuits which use two transistors connected together in such a way that they are always in opposite states to each other. Another term for these circuits is **flip-flop**. There are three types of multivibrator: **bistable, monostable** and **astable**.

Bistable multivibrator

The word 'bistable' means that it is stable in either of two states. The basic circuit is shown in Figure 18.1.

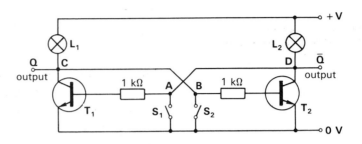

Figure 18.1 Simple bistable multi-vibrator or 'flip-flop'

This is really two transistor switches connected back-to-back. To make the diagram neater the transistor switch around T_1 has been drawn 'back-to-front'. Of course it still behaves in the same way. Two outputs are shown, Q and $\bar{Q}$, and we shall see that these outputs will always be in opposite states to each other.

How the simple bistable behaves

The bistable behaves like a **latch**. Suppose that we turn T_1 off by closing S_1. Then $V_C = +9V$. This turns T_2 on and so $V_D = 0V$. This will keep T_1 off, even if we open S_1 again. The circuit has latched and is **stable** in this state. The opposite happens if we close S_2 instead. T_2 will turn off and $V_D = +9V$. This will turn T_1 on and make $V_C = 0V$ even if we open S_2 again. Once again the circuit has latched and is stable. The bistable 'remembers' which was the last switch to be closed. In fact a circuit like a bistable is used in a **computer memory**.

Bistable latches are used to hold digital displays steady so that they can be read easily.

Monostable multivibrator

The circuit is similar to the bistable, but with a capacitor on the base of T_1 (Figure 18.2). Its stable state is with the lamp L_2 off (and L_1 on). If the switch S is pressed and released this triggers the monostable. Lamp L_2 goes on for a period then goes off again.

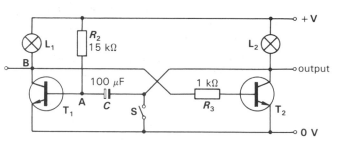

Figure 18.2 Monostable multivibrator

How it works

If we close S we will turn T_1 off. V_B goes to 5 V and T_2 will be turned on. When S is opened, then C will charge up through R_2 so V_A rises. After a while V_A will pass 1 V and T_1 will turn on and T_2 turn off again.

Changing the monostable period

With the component values shown the monostable period is about 1 second. This may be increased by increasing the value of C or R_2.

Astable multivibrator

The circuit is like two monostables connected together. (See Figure 18.3.) It is unstable in either state. The lamps will keep flashing on and off. If R_2 and R_3 as well as C_1 and C_2 are equal then the lamps flash evenly. The mark-to-space ratio is 1:1.

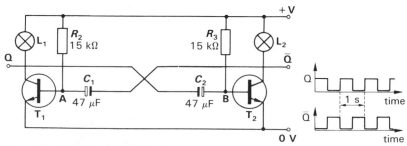

Figure 18.3 Astable multivibrator

How it works

In effect we have two monostables always in opposite states to each other. When the unstable one drops back to its stable state it will force the other one into an unstable state and so on.

A monostable can be used to control a photographic enlarger. When the button is pressed, the lamp goes on for a preset length of time which can be adjusted by a variable resistor.

R_2 and R_3 (kΩ)	C_1 and C_2 (μF)	frequency (Hz)
15	47	1
15	10	5
1	4.7	150
1.5	0.47	1000

Figure 18.4 Typical frequencies of the astable multivibrator with various values of R and C

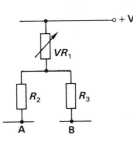

Figure 18.5 Connecting a variable resistor into the astable multivibrator

Changing the frequency

With the component values shown the astable frequency is about 1 Hz. The frequency can be increased by decreasing the values of R_2 and R_3 or C_1 and C_2. Some examples are shown in Figure 18.4.

Variable frequency

The simplest way to vary the frequency is to use a variable resistor as shown in Figure 18.5.

Uses of an astable multivibrator

We have only used lamps on T_1 and T_2 so that it is easy to see what is happening. Like other astables this circuit can be adapted to a number of different uses by changing the component values and connecting different components in place of the lamps. For example, we can replace one lamp with a resistor and the other with a speaker to make sounds, as in Figure 18.6. Depending on the values of the components these can be 'clicks' (like a metronome or a clock) or a musical note.

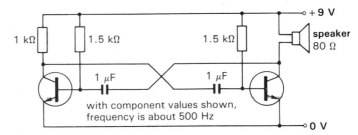

Figure 18.6 Using an astable multivibrator to produce a note

Unequal mark-to-space ratio

Look again at Figure 18.3. If we make R_3 larger than R_2 then T_2 will take a longer time to switch on than T_1. The mark-to-space ratio is unequal.

In the circuit shown in Figure 18.7 the mark-to-space ratio is about 1:10. One use for this might be to turn a car's windscreen wipers on once every 10 seconds. (Useful in drizzle!)

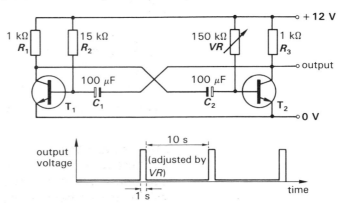

Figure 18.7 Astable with a 1:10 mark-to-space ratio

47

Summary

- ● Multivibrator circuits use two transistors.
- ● A bistable multivibrator has two stable states. It can be used as a latch or as a memory element in computers.
- ● A monostable multivibrator uses one capacitor.
- ● An astable multivibrator uses two capacitors.
- ● The frequency of the astable is increased if the capacitors or timing resistors are decreased in value.
- ● If the capacitors have equal values then the mark-to-space ratio is equal (1:1).

Questions

1 Draw the bistable multivibrator circuit shown in Figure 18.1.
2 Copy the following and fill in the missing words (e.g. high, low, on, off):
   'When switch S_1 is closed, V_A goes ____ and T_1 is turned ____. This makes V_C ____ and because the base of T_2 is connected to point C this turns T_2 ____. This makes V_D go ____ and this keeps point A ____, even if switch S_1 is opened again.'.
3 What two ways are there of making the monostable period of a monostable multivibrator shorter?
4 Look at the circuit diagram in Figure 18.8.

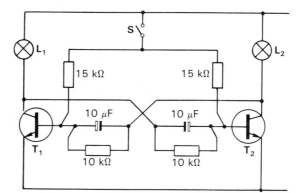

Figure 18.8

When S is closed the circuit is an astable with a frequency of 5 Hz. When S is opened the circuit becomes a bistable.
(a) Describe what happens to the lamps when S is closed for a while and then opened.
(b) Suggest a possible use for this circuit.

19
Exercises

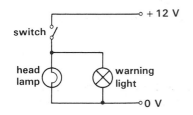

Figure 19.1

These questions are all about designing a system for use in a car to let the driver know when the headlights are switched on.

1   The simplest way of doing this is shown in Figure 19.1. Does this system tell the driver if the lights are actually working? Explain your answer.

2   Now look at the system in Figure 19.2.
(a) Will the output from the light sensor be a high voltage or a low voltage when it is in the dark?
(b) Will the warning light be on or off when the LDR is in the dark?
(c) How could this system be used to let a driver know if the headlights were working or not?
(d) Where would each part of the system be put in the car?
(e) Would the system be connected through its own on/off switch to the battery or would it use the same switch as the headlights? Explain your answer.

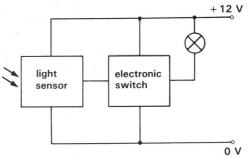

Figure 19.2

3   Look at the circuit diagram in Figure 19.3.
(a) Describe what happens to the voltage at A and at C when the LDR goes from being in the light to being in the dark.
(b) Why is this a better warning system than the previous one?

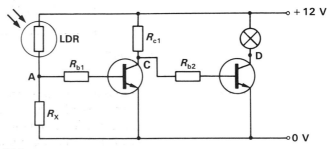

Figure 19.3

4   Suppose that we wanted to make the warning light **flash**:
(a) What is the name of the circuit which could do this?
(b) Draw a block diagram of the new complete system.
(c) Why might a flashing light be better than a steady light?
(d) Suggest a suitable frequency at which to flash the light.

49

5   (a) The light dependent resistor (LDR) used in the circuit in Figure 19.4 changes its resistance from 100 kΩ in the dark to 1 kΩ in a bright light.

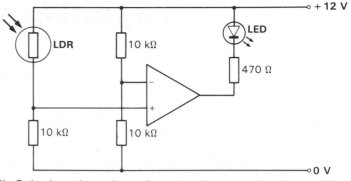

Figure 19.4

(i) Calculate the voltage input to the operational amplifier (op-amp) at the inverting input.   (ii) Calculate the voltage input to the op-amp at the non-inverting input when the LDR is in the dark.   (iii) Calculate the voltage input to the op-amp at the non-inverting input when the LDR is brightly lit.

(b) The op-amp (power connections not shown) has its output connected to a light-emitting diode (LED).   (i) Will the LED light when the LDR is in the dark or in the light?   (ii) Explain why the LED lights when the LDR is in only one of these situations.

(c) This circuit can be adapted for a car so that the driver, sitting in the normal position, could test all four of the car's main lights before driving off.   (i) Draw a circuit for testing all of the car's main lights.   (ii) State where, in the car, various parts of your circuit would have to be fitted.   (iii) Describe how the device would be used.

(d) Draw an addition to the circuit you have suggested that would enable the driver to adjust the point at which the LED would begin to operate.   (*AEB 85 paper 2*)

6   Figure 19.5 shows a further addition to the system.
    (a) What will be heard from the speaker?
    (b) How might this arrangement be used in the system?
    (c) Draw a block diagram to show how it can be included.

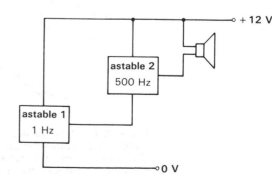

Figure 19.5

50

Section 3

Digital Systems

20
Analogue and Digital Signals

In an analogue signal the voltage will generally vary over the full range possible, from 0 V up to the voltage of the power supply. A typical analogue signal is shown in Figure 20.1. The audio signal (chapter 30) is an **analogue signal**.

A digital signal (Figure 20.2) is much simpler and only two voltages are important. These are called **logic levels**. A low voltage, i.e. around 0 V, is called a **logic 0**; a high voltage, i.e. more than 1 V, is called a **logic 1**.

The input and output signals from electronic switches (section 2) are really digital (either on or off). When a light pen reads a bar code it produces a digital signal which is similar to the one in Figure 20.2. This is fed into a computer. See Figure 20.3.

Figure 20.1   A typical analogue signal

Sending digital signals

Although digital signals are very simple, they can be used to send very complicated information provided we have some way of understanding them. This is like Morse code in which messages can be sent by a combination of dots and dashes only.

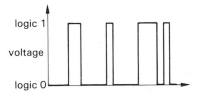

Figure 20.2   A typical digital signal

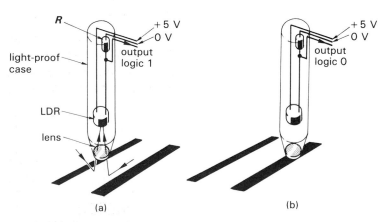

Figure 20.3   A bar code is a common sight on food packets etc. It is 'read' by a probe similar to the one shown here.   (a) A white surface reflects light back to the LDR and the probe produces a high voltage (logic 1)   (b) A black surface does not reflect so the probe produces a low voltage (logic 0).

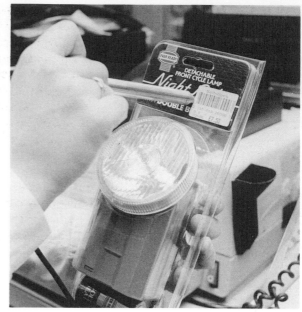

Sending and receiving Morse code needs much skill and practice. It was originally developed for use in the telegraph system before being adapted for radio transmission.

The bar code is like a modern version of Morse code. Instead of dots and dashes, high and low voltages are produced. What information might be contained in this bar code?

Unfortunately it takes a skilled operator to convert words into Morse code and then convert the received message back into words. In modern electronic systems the information is converted to digital form, transmitted, processed and turned back into a form that we can understand by electronic circuits. This section deals with these circuits and how they behave.

Why use digital signals?

Although analogue signals are quite easy to produce (e.g. a microphone converts sound waves into an analogue audio signal) they can quite easily become distorted. For example an amplifier might **clip** the signals. If they are sent over long distances, **noise** may interfere with the signal and so on.

The main advantage of sending information by digital signals is that there is much less chance of the signal getting distorted or scrambled on the way. (This was why Morse code was used originally. In the early days of radio the signals were so weak that speech could easily become lost in all the background noise, but even faint dots and dashes could be picked out of the 'hiss'.)

A low voltage can only be turned into a high voltage accidentally by a very large amount of interference. Also, clipping is not a problem.

Digital signals in computers

Another use for digital signals is in computers. By simply opening and closing electronic switches, information in the form of **logic 1** and **logic 0** can be sent round the system. It can be made to do different things depending on whether a '1' or a '0' is received. We can do arithmetic if numbers are converted to **binary** (chapter 23) and pictures can be produced on a screen by 'telling' the screen to 'put a dot' (1) or 'do not put a dot' (0) at various points. To do all this with analogue signals would be unreliable and much slower. Some things would be impossible.

Summary

- In **analogue signals** voltages vary continuously over the full range from 0 V to the power supply voltage.
- In **digital signals** only two voltage levels are important: 0 V is called logic 0, and more than about 1 V is logic 1.
- An analogue system is less reliable and more prone to noise than a digital system.
- A digital system behaves according to simple rules of logic.

Questions

1. Describe the difference between analogue and digital signals.
2. Why is a digital signal less likely to be distorted?
3. Which of these give analogue and which give digital signals?
   (a) a record pick-up   (b) a transistor switch
   (c) a sine wave oscillator   (d) an astable
   (e) a monostable

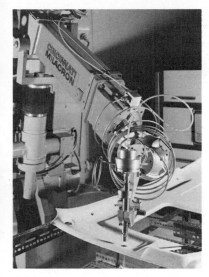

A robot arm, seen here cutting out fibreglass for car headlamps, is controlled with great precision by digital signals.

21
Logic Gates

Logic gates are one of the ways that digital signals can be controlled. They use very simple rules to produce an output which depends on the signals going into their inputs.

Truth tables
The easiest way of describing how each gate behaves is to draw up a truth table. This shows all the possible combinations of input signals and the output signal produced in each case.

OR gate
The OR gate shown in Figure 21.1 has two inputs. The output C is '1' if input A or input B receives a logic 1 signal.

An OR gate could be used to allow a TV to change channels by a switch on the set or remote control (Figure 21.2).

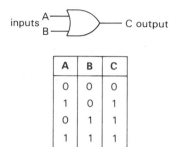

A	B	C
0	0	0
1	0	1
0	1	1
1	1	1

Figure 21.1 An OR gate and its truth table

Figure 21.2 Typical use of an OR gate. Here it is being used to allow a TV set to change channels by a switch on the set or by a signal from the remote control.

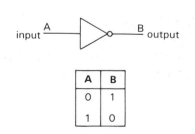

The inverter (or NOT gate)
This is the simplest gate. The symbol and the truth table are shown in Figure 21.3. Notice that the output is **not** the same as the input. The inverter has already been discussed in chapter 10.

NOR gate
The NOR gate (Figure 21.4) behaves like an OR gate followed by a NOT gate. The output C is only '1' if neither input A nor input B receives a logic 1 signal.

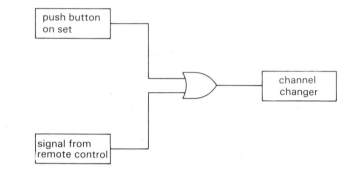

A	B
0	1
1	0

Figure 21.3 A NOT gate or INVERTER and its truth table

A	B	C
0	0	1
1	0	0
0	1	0
1	1	0

Figure 21.4 A NOR gate and its truth table

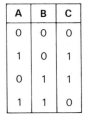

A	B	C
0	0	0
1	0	1
0	1	1
1	1	0

Figure 21.5 An XOR gate and its truth table

Exclusive OR (XOR) gate

The XOR gate shown in Figure 21.5 has two inputs. The output C is '1' if only input A or only input B receives a logic 1 signal. (When A and B together are 1, the output C is a 0.)

The XOR gate could be used to allow two switches to control one light (Figure 21.6).

Either switch can turn the lamp on, then either switch can turn it off again.

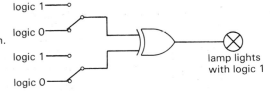

Figure 21.6

inputs A B — C output

A	B	C
0	0	0
1	0	0
0	1	0
1	1	1

Figure 21.7 An AND gate and its truth table

AND gate

The AND gate shown in Figure 21.7 has two inputs. The output C is only '1' if both input A and input B receive a logic 1 signal.

An AND gate could be used to make sure that a washing-machine only fills with water if the programme is set **and** the door is closed:

A B — C

A	B	C
0	0	1
1	0	1
0	1	1
1	1	0

Figure 21.9 A NAND gate and its truth table

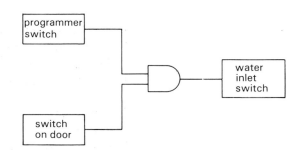

Figure 21.8 A typical use of an AND gate is to make sure that a washing-machine will only fill with water if the programme is set and the door is closed

NAND gate

The NAND gate (Figure 21.9) behaves like an AND gate followed by a NOT gate. The output C is only **not** '1' if input A and input B receive a logic 1 signal.

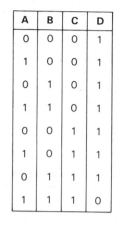

A	B	C	D
0	0	0	1
1	0	0	1
0	1	0	1
1	1	0	1
0	0	1	1
1	0	1	1
0	1	1	1
1	1	1	0

Figure 21.10 A 3-input NAND gate and its truth table

More inputs

All of the gates (except the NOT gate) could have more than two inputs. Although the truth table will be quite long, it still follows the same rules of logic. For example a three-input NAND gate is shown in Figure 21.10.

Connecting inputs together

If we connect the two inputs of a NOR gate together, we produce the following truth table (Figure 21.11):

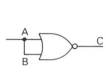
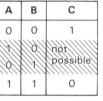

A	B	C
0	0	1
1	0	not possible
0	1	not possible
1	1	0

A	C
0	1
1	0

Figure 21.11 A NOR gate behaves like a NOT gate if its inputs are connected together

The gate now becomes a simple NOT gate. A similar thing will happen if we connect the two inputs of a NAND gate together.

Integrated circuit gates

All of the gates can be bought as integrated circuits. These are quite cheap (from about 10p) and usually one 'chip' contains several identical gates in one package, all sharing the same power supply. Figure 21.12 shows two common examples. The gates can simply be connected together by short wires or copper tracks. This means that very complex circuits can be built on a small circuit board.

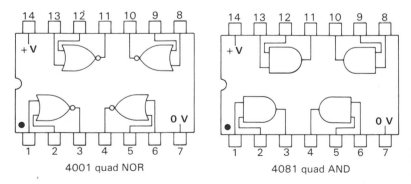

Figure 21.12 Pinouts of two common IC gate packages

4001 quad NOR   4081 quad AND

Summary
- The output from a logic gate depends on the logic levels put on the inputs according to simple rules.
- A truth table is a chart showing all the possible inputs and the corresponding output for each combination.
- The six types of gates are NOT, OR, AND, XOR, NOR, NAND.
- Gates can be connected together and their behaviour worked out from the truth table.
- Integrated circuit gates are compact and easy to use.

Questions
1 Draw the truth tables and symbols for the six types of gate.
2 Draw a truth table for:
   (a) a three-input OR gate   (b) a three-input NOR gate.
3 Suggest a use for:
   (a) a three-input AND gate   (b) a two-input NOR gate.
4 Figure 21.13 shows a combination of gates. Draw a truth table to show how it behaves.
5 A safety feature on a recent gas water heater is an alarm which is triggered either if the flame goes out or if the temperature gets too high. Two of the system's building blocks (and truth tables) are:

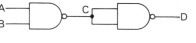

Figure 21.13

flame detector	Flame on	Flame off
output	1	0

temperature sensor	OK	Too hot
output	0	1

The alarm needs a logic 1 on its input to operate. The other building blocks are a NOT gate and an AND gate. Draw a block diagram of how the system is organised.

22
Combining Logic Gates

We shall often meet systems in which several logic gates have been combined together. In this chapter we shall look at some typical examples and work out the way the combination behaves.

Drawing up a truth table

Start by labelling all the different points of the circuit including the inputs and output. Then draw up a truth table of all the points and fill in all the possible combinations of inputs.

Example

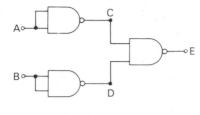

A	B	C	D	E
0	0			
1	0			
0	1			
1	1			

circuit   truth table

Fill in the truth table in stages like this:

A	B	C	D	E
0	0	1	1	
1	0	0	1	
0	1	1	0	
1	1	0	0	

stage 1

A	B	C	D	E
0	0	1	1	0
1	0	0	1	1
0	1	1	0	1
1	1	0	0	1

stage 2

A	B	E
0	0	0
1	0	1
0	1	1
1	1	1

Figure 22.1 Simplified truth table

In Figure 22.1 we have just shown the inputs and the output. We can see that the combination behaves like a simple OR gate. In fact this circuit is a way of making an OR gate from a combination of NAND gates.

A more complicated system may need several stages before you finish, but you can still get the right answer if you work in this systematic way.

Now let us look at a more complicated system:

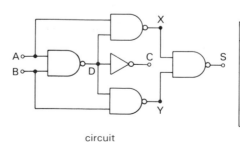

A	B	D	X	Y	C	S
0	0	1				
1	0	1				
0	1	1				
1	1	0				

circuit   truth table

A	B	D	X	Y	C	S
0	0	1	1	1		
1	0	1	0	1		
0	1	1	1	0		
1	1	0	1	1		

next stage

A	B	D	X	Y	C	S
0	0	1	1	1	0	0
1	0	1	0	1	0	1
0	1	1	1	0	0	1
1	1	0	1	1	1	0

last stage

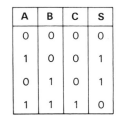

A	B	C	S
0	0	0	0
1	0	0	1
0	1	0	1
1	1	1	0

Figure 22.2 Simplified truth table

Figure 22.2 shows what the truth table looks like if we just look at the outputs and the inputs. Notice that we have labelled the outputs S and C. There is a reason for this, as we shall see in the next chapter.

Summary
● Combinations of logic gates can be worked out if you draw up a truth table showing each point in the circuit.
● It is best to fill in the truth table in stages, starting at the inputs.

Questions
1   Look at the combinations of gates in Figure 22.3 and work out their truth tables.

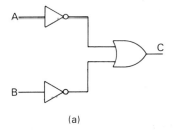

(a)

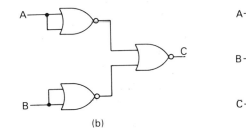

(b)

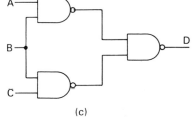

(c)

Figure 22.3

23
Binary Arithmetic

It was mentioned in chapter 20 that computers can use digital signals to do arithmetic by working in **binary**. First of all let us look at the **decimal** and **binary** system of numbers and then see how logic gates can be used to do calculations in binary.

The decimal system

The numbers we are familiar with are part of the **decimal system** of counting. Ten symbols are used in all: 0, 1, 2, 3, 4, 5, 6, 7, 8, 9. We write decimal numbers as being so many **units**, **tens**, etc., e.g. 432 means **4 hundreds, 3 tens and 2 units**.

The binary system

In the **binary system** only two symbols are used: 0 and 1. We write binary numbers as being so many **ones**, **twos**, **fours**, etc., e.g. '101' means **1 four, 0 twos and 1 one** (i.e. '5' in decimal). The first sixteen numbers are shown in both systems in Figure 23.1.

binary				decimal	
eights	fours	twos	ones	tens	units
0	0	0	0	0	0
0	0	0	1	0	1
0	0	1	0	0	2
0	0	1	1	0	3
0	1	0	0	0	4
0	1	0	1	0	5
0	1	1	0	0	6
0	1	1	1	0	7
1	0	0	0	0	8
1	0	0	1	0	9
1	0	1	0	1	0
1	0	1	1	1	1
1	1	0	0	1	2
1	1	0	1	1	3
1	1	1	0	1	4
1	1	1	1	1	5

Figure 23.1 The first sixteen numbers in binary and decimal form

Writing binary numbers

It is often useful to label each digit of a binary number, starting from the right. An example is given in Figure 23.2, where 'A' = the binary number '1001'.

$$A = \begin{array}{|c|c|c|c|} \hline A_3 & A_2 & A_1 & A_0 \\ \hline 1 & 0 & 0 & 1 \\ \hline \end{array}$$

Figure 23.2 Note that we start counting from zero, so the third digit, for example, is called the A_2 digit

Binary arithmetic

To add two binary numbers we use the following rules:

$$0 + 0 = 0$$
$$0 + 1 = 1$$
$$1 + 1 = 0 \text{ carry } 1$$

e.g:

110	(6 in decimal)
011 +	(3 in decimal)

1001 (sum) (9 in decimal)

11 (carry)

Binary and logic gates

The simple rules of binary arithmetic can be copied by logic gates (Figure 23.3):

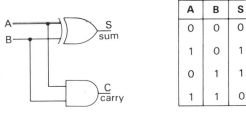

A	B	S	C
0	0	0	0
1	0	1	0
0	1	1	0
1	1	0	1

Figure 23.3

The XOR will add two binary digits together to give their sum. The carry digit is only 1 when the two digits being added are 1, so an AND gate can give the carry digit.

This combination of gates is called a **half adder**. It will add two binary digits and produce the sum and carry digit. We saw another way of producing the same result in chapter 22. A half adder is so useful that it is given its own symbol (Figure 23.4).

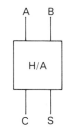

Figure 23.4 Half adder

Full adder

Full addition needs to be able to add together three digits (the two digits being added and any carry digit). This can be done by the system shown in Figure 23.5. It is called a **full adder**.

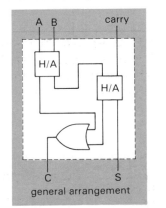

general arrangement

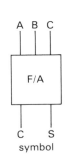

symbol

Figure 23.5 Full adder

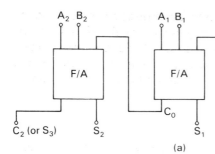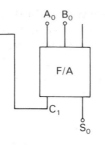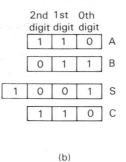

(a)

(b)

Figure 23.6 (a) This 3-bit adder will add together two numbers each with 3 digits (b) Example of binary addition

By connecting full adders together in a chain, longer binary numbers can be added (Figure 23.6).

Other arithmetic

Multiplication is carried out by repeated addition. Although this takes a lot of steps, a computer can work very quickly! It is also possible to carry out subtraction and division (which is really repeated subtraction). The actual methods used vary and will not be discussed here.

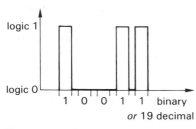

Figure 23.7 One method of representing a number by a digital signal

Binary numbers and signals

Since both binary numbers and digital signals use only two different states (1 or 0), we could use a digital signal to stand for a binary number. See Figure 23.7.

Summary
● Binary arithmetic uses two digits, 1 and 0, and follows simple rules.
● The rules can be copied by logic gates which can be used to carry out binary arithmetic.
● Binary numbers can be transmitted by digital signals.

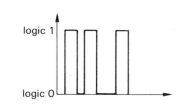

Figure 23.8

Questions
1 Write '12 + 7' in binary form and work out the binary sum.
2 If you wanted to add up the binary numbers 0111 and 1001 how many full adders would you need? Draw a diagram.
3 What number might the digital signal in Figure 23.8 represent?

24
Bistable Latches

In the last chapter we saw how binary arithmetic can be carried out by logic gates. This, of course, is only possible if the signals can be put on the inputs and **held** there while the calculation is carried out. It is also useful to have a way of holding the result or other binary information until it can be used. Circuits which can store binary information are **latches**.

Bistable latches

Since only two kinds of information need to be stored (1 or 0) a **bistable** (which has two possible **states**) is ideal. There are several different types of bistables. Two types are the D-type (data type) and the RS-type (reset, set type) latches.

The D-type bistable latch

This type of bistable has two inputs; the **data** input (D) and the **clock** input (C) (Figure 24.1).

The bistable can only change state when a signal which goes from logic 1 to logic 0 is sent into C. (See Figure 24.2.) When this happens the output Q goes to the same logic level that is on the D input. The other output, $\overline{Q}$, is always at the opposite level to Q.

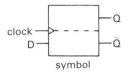

Figure 24.1 D-type latch

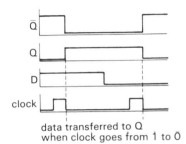

Figure 24.2 Timing diagram for a D-type latch

Registers

A number of D-type latches could be used to store the result of a binary calculation (Figure 24.3):

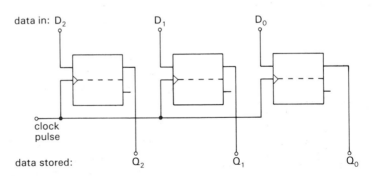

Figure 24.3 This register can store three different 'bits' of data. It is called a 3-bit data register.

This arrangement is called a **register**. The number is stored by sending in a **clock pulse**. A register like this is used to hold the output for a **digital display** steady so that it can be read.

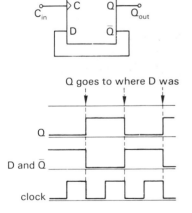

Figure 24.4 Dividing the frequency of a square wave by two using a D-type latch

Dividing the frequency of a square wave by two

A more common use of a D-type latch is shown in Figure 24.4. A square wave is sent into the C input. A square wave is also produced at the Q output, but at **half the frequency**.

Counting using D-type latches

In Figure 24.5 we connect three D-type latches in a chain and feed a series of square wave pulses into the C input of the first. It is easy to see from the timing diagram that the frequency is being constantly divided by two along the chain.

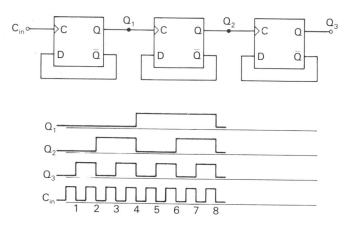

Figure 24.5   3-bit binary counter

Q_3	Q_2	Q_1	C_{in}
0	0	0	0
0	0	1	1
0	1	0	2
0	1	1	3
1	0	0	4
1	0	1	5
1	1	0	6
1	1	1	7

Figure 24.6   Table of results from the 3-bit binary counter

Figure 24.6 puts the results of seven clock pulses in a table. If we look at these results we can see a familiar pattern; the first eight binary numbers! The arrangement of D-type bistables shown in Figure 24.5 is a **binary counter**. You should be able to see that the next clock pulse will produce the output 000. This counter can only go up to 7 (in decimal) before it starts again. Higher numbers can be counted by using more bistables.

Circuit for a D-type bistable

A D-type bistable can be built from a bistable multivibrator together with some extra components (see chapter 18) but, not surprisingly, it is easier and cheaper to use integrated circuits. The 4013 contains two D-type latches and is shown in Figure 24.7.

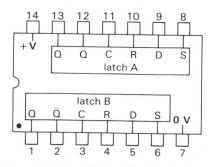

Note, the S and R inputs let us set (to 1) or reset (to 0) the Q of each latch, independently of D or C pulses.

Figure 24.7   Pinout of a 4013 dual D-type latch

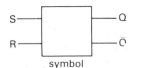

Figure 24.8 An RS-type latch

The RS-type latch

This type of latch is shown in Figure 24.8. It has two inputs, S (or **set**) and R (or **reset**), and two outputs (once again Q and $\overline{Q}$) which are always in opposite states to each other. The truth table is shown in Figure 24.9 and perhaps needs some explanation.

If both R and S are at logic 1 then the outputs stay as they are. If only S is taken to logic 0 then output Q becomes 0. If only R is taken to logic 0 then output Q becomes 1. Either of these two states can be stored by making both R and S logic 1 again.

Notice that if both R and S are at logic 0 the output could be anything. This might even damage the latch and is not allowed.

R	S	Q	$\overline{Q}$	
0	0			not allowed
0	1	1	0	Q set to 1
1	1	1	0	stores last state
1	0	0	1	Q reset to 0
1	1	0	1	stores last state

Figure 24.9 Truth table for an RS-type latch

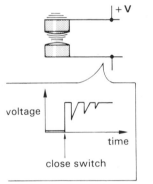

Figure 24.10 Effect of contact bounce in a switch. Instead of a clean change from logic 0 to logic 1, a series of pulses is produced.

Uses of RS latches

A common use of an RS latch is to **clean up** the action of a mechanical switch when it is being used in digital systems. When a switch is operated its contacts tend to 'bounce' slightly so that it may in fact make and break contact several times before it settles. (See Figure 24.10.) Although this only takes a fraction of a second it could cause problems in digital systems by sending several pulses when only one was intended. An RS latch can be used to **debounce** the switch. (See Figure 24.11.)

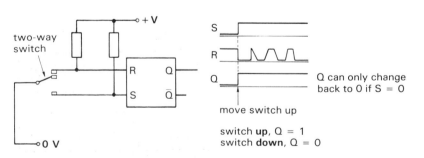
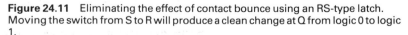

Figure 24.11 Eliminating the effect of contact bounce using an RS-type latch. Moving the switch from S to R will produce a clean change at Q from logic 0 to logic 1.

65

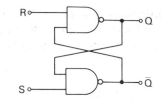

Figure 24.12 RS-type latch using cross-coupled NAND gates

Circuit for an RS latch

An RS latch can be made by connecting two NAND gates together as shown in Figure 24.12. A truth table can be drawn to show how this behaves. This is left as an exercise at the end of this chapter. An IC containing RS latches is the 4043 shown in Figure 24.13.

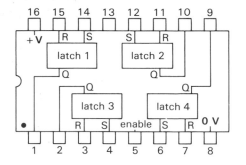

Figure 24.13 Pinout of a 4043 quad RS-type latch

Summary
● A latch stores a binary digit (1 or 0).
● A D-type latch stores the data input level when it receives a clock pulse.
● A D-type latch can be used in registers to hold data in a binary counter or to halve the frequency of a square wave.
● An RS-type latch stores data when both inputs are logic 1.
● An RS-type latch can be used to debounce a switch.

Questions
1 Draw the symbols for a D-type and an RS-type latch and explain how each one can be made to store binary information.
2 Look at the arrangement of D-type bistables in Figure 24.14 and write down:
 (a) the highest number it can count to
 (b) the logic levels of each Q after 12 clock pulses
 (c) the frequency of the signal at Q_4 if the clock pulse is 4 kHz.
3 Draw a truth table and a timing diagram for the two NAND gates shown in Figure 24.12 to show what happens if:
 (a) Q starts at logic 1 while R and S are both at logic 1
 (b) R is then taken to logic 0
 (c) R is taken back to logic 1.

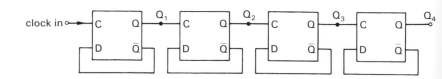

Figure 24.14

4   At show jumping events an electronic clock is started when the horse and rider first pass through a beam of light. The clock stops when they pass through the beam again. The system contains each of the following building blocks:

light source

light sensor (which gives logic 1 when receiving light)

an electronic clock (which runs when its input is at logic 1)

a bistable (which changes state when its input goes from 1 to 0).

(a) Draw a block diagram to show how these may be connected together to make the automatic timing system described.

(b) Draw a timing diagram of how the signals behave.

(c) When the system was first installed it was found that sometimes the beam was broken twice in quick succession; first by the horse's head then again by the rider. What problems might this cause?

(d) How would a monostable (or 'triggered pulse producer') help to make the system more reliable? Suggest a suitable monostable period in this case.

25
Counters and Displays

Suppose that we have a stream of pulses coming from an astable. We may want to **count** the number of pulses over a period of time and then **display** the result. This chapter shows how this can be done.

How high can we count?
The highest number that a binary counter can go up to before repeating itself depends on the number of **bits** in the counter. A **4-bit counter** could count up to binary '1111' (i.e. '15' decimal). We could count higher by using more bits in our counter, but it is better to use a system called **binary-coded decimal**.

Binary-coded decimal
The number '927' could be written as '1001 0010 0111'. We have taken each digit in the decimal number and put it into its binary form. This is called **binary-coded decimal** or **BCD**.

Counting in binary-coded decimal
A different counter is used for each decimal digit (i.e. one counter for the units, one for the tens, one for the hundreds, etc.). Figure 25.1 shows how this can be done.

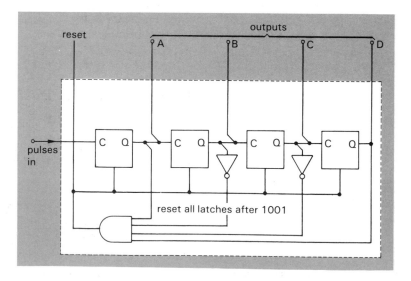

Figure 25.1 A 4-bit coder for binary-coded decimal (BCD) counting

This 4-bit counter can only count from 0 up to 9. It is automatically reset to '0000' after it has counted to '1001' ('9' in decimal). You can check this by drawing up a truth table for the gates.

The four output lines are labelled A, B, C, D.

Counting higher

Higher numbers can be counted by using more BCD counters and **cascading** them together so that after ten clock pulses the units counter sends a pulse to the tens counter moving it on by one and so on. Figure 25.2 shows the principle of this. Each BCD counter has four outputs (one for each binary digit).

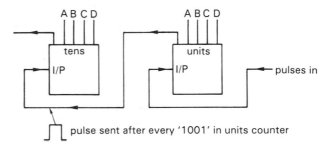

Figure 25.2 Cascaded BCD counters

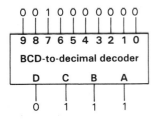

Figure 25.3 BCD-to-decimal decoder

Decoders

The information from the BCD counter is still in binary form. If we want to turn it back into decimal we will need a **binary-to-decimal decoder**. This has four inputs and ten outputs (see Figure 25.3).

Only one output line at a time is at logic 1. The others are at logic 0. Each output line goes to logic 1 when its particular code appears on the four input lines. Figure 25.3 shows '0111' on the output and therefore output '7' at logic 1.

How a decoder works

Inside the decoder is a network of logic gates. Figure 25.4 shows the gates used to decode '0111' into '7'. Each output has its own set of gates. There is no need to actually make a decoder; there are ICs which can do the whole job for us (Figure 25.5).

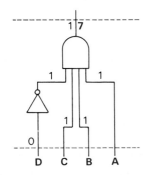

Figure 25.4 Part of the gate network inside a BCD-to-decimal decoder

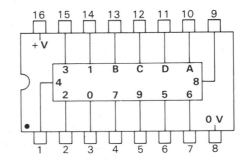

Figure 25.5 Part of a 4028 BCD-to-decimal decoder IC

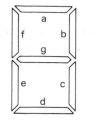

Figure 25.6 A 7-segment display

Displays

Now that we have counted our pulses we must display them in a way that we humans can understand. A common way of displaying numbers (and some letters) is a **7-segment display**. (See Figure 25.6.) By lighting up some of the segments the full range of numbers can be displayed (Figure 25.7):

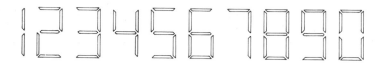

Figure 25.7 Numerals on a 7-segment display

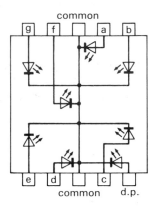

Figure 25.8 Pinout of a typical 7-segment LED display

Types of displays

LED displays use seven **light-emitting diodes**. One end of each LED is joined to a **common** line. The other end goes to one of the input pins. (See Figure 25.8.) Often an eighth LED is used as a decimal point. LED displays can be seen easily but they take a lot of current.

LCD displays use seven **liquid crystals**. They are connected in the same way as the LED display. When a voltage is put on a liquid crystal it goes black, making it stand out. Although they use less current than LED displays, LCD displays cannot be seen in the dark.

Decoding for displays

In order to make the number '7', segments a, b, c only must be lit up. The number '6' requires segments a, f, g, c, d, e. A **diode network** is used to make sure that each number lights up only the segments it needs. Figure 25.9 shows the matrix for the inputs '6' and '7'.

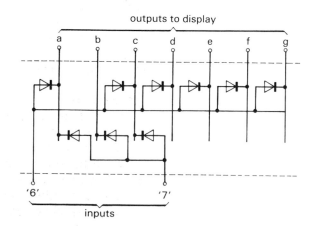

Figure 25.9 Part of a diode matrix for a decimal-to-7 segment decoder

Block diagram of the system

The block diagram in Figure 25.10 shows how we can count up to ten pulses (0 to 9) and display the result. Remember, if we want to count higher we can use another decade counter and display for the tens and so on.

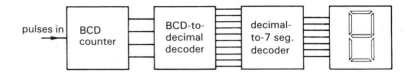

Figure 25.10 Block diagram of a counter and display. The first three boxes make a decade counter.

BCD-to-7 segment decoders

Not surprisingly there are ICs which do the complete process from taking a 4-bit BCD number to producing the signals for a 7-segment display. This is the easiest way of counting pulses and displaying the result. (See Figure 25.11.)

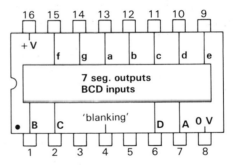

Figure 25.11 Pinout of a 4511 BCD-to-7 segment decoder

Summary

● Counting is usually done in binary-coded decimal using 4 **bits** to encode the numbers from 0 to 9.
● Binary-to-decimal decoders take the output from a 4-bit BCD counter and convert this to one of ten decimal outputs.
● Numbers are displayed with a 7-segment display which need their own type of decoder.
● LED displays are bright, but use more current than LCD types.

Questions

1   What do the following stand for?
    (a) BCD   (b) LED   (c) LCD
2   If a binary-to-decimal converter only counts up to 9, why does it have 10 output lines?
3   What is the BCD number '1010 1100 0110' in decimal?
4   If this number were displayed using three 7-segment LED displays, which segments of each display would be lit up?

71

26
Sequential Systems

In this chapter we will see how logic gates can be combined into larger building blocks to make sequential systems.

Systems in which events happen one after another are called **sequential systems**. Traffic lights are a good example of this.

Block diagram of traffic lights

The block diagram in Figure 26.1 shows one way of producing the order of lights.

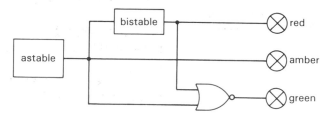

Figure 26.1   Block diagram of a traffic light system

The easiest way to show what happens is to draw timing diagrams (Figure 26.2):

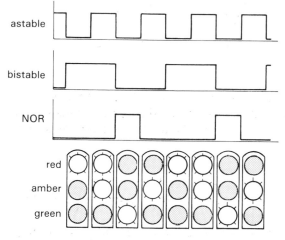

The astable produces square waves. These feed the bistable and amber light.

The bistable changes state when its input goes from 0 to 1.

The NOR gate has a 1 output only if neither the astable nor the bistable output is 1.

The red light follows the bistable.
The amber light follows the astable.
The green light follows the NOR gate.

Figure 26.2   Timing diagram for a traffic light system

You can see that this produces the correct sequence of lights.

A video recorder goes through a precise sequence of operations. Some of these are controlled by a digital system.

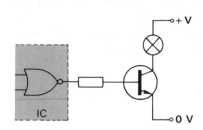

Figure 26.3 Using a transistor as a
buffer between an IC and a lamp

Controlling the lights

If we tried to connect a lamp straight to the output of an IC logic
gate, too much current would flow through the gate and it would
be destroyed.

Transistor buffer

Look at Figure 26.3. The output from the NOR gate is fed into a
transistor switch which can control a larger current. It is used as a
buffer (or **interface**) between the IC and the lamp.

More current

If we want to drive an even larger current (say for a motor) we can
always connect a relay to the output of the transistor. See Figure
26.4. (Note the **protection diode**—see chapter 79.) We must not
connect the relay directly to an IC gate.

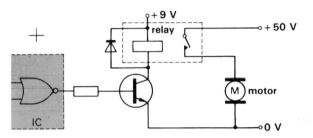

Figure 26.4 Driving a relay to control a
large load

Summary

● Logic gates can be combined into useful sequential systems.
● It is easiest to work out how a sequential system behaves by
  drawing a timing diagram and working through from the input.

Questions

1 What is meant by a sequential system?
2 What is the best way to work out how a sequential system
   behaves?
3 Look at the system in Figure 26.5 and complete the timing
   diagram to show how it behaves. (Remember that an AND gate
   only produces a logic 1 when both inputs are at logic 1.)

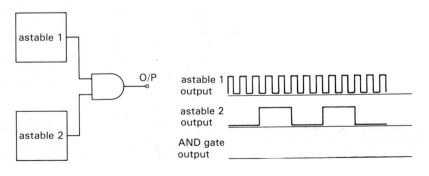

Figure 26.5

27
Analogue to Digital

So far we have produced digital signals from digital circuits. Sometimes we want to change an analogue signal into a digital signal. This needs an **analogue-to-digital** (or **A to D**) **converter**.

An A to D converter. Notice the ribbon cable at the top. Each wire carries one 'bit' of digital information.

A typical digital voltmeter.

Digital voltmeter

A digital voltmeter (DVM) measures voltage and displays the result on 7-segment displays. They use an A to D converter to turn an analogue signal (the voltage to be measured) into a digital signal suitable for driving the display.

Block diagram

Figure 27.1 shows one type of DVM. Its A to D converter works by counting pulses.

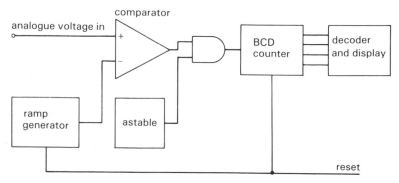

Figure 27.1   Block diagram of a digital voltmeter (DVM)

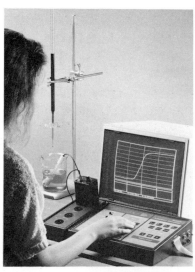

This scientist is measuring the pH of a liquid. The pH probe produces an analogue voltage. This is converted to a digital signal and can be stored for display later.

How it works

A timing diagram shows what happens (Figure 27.2). Suppose that the counter and ramp generator have just been reset to zero. Now:

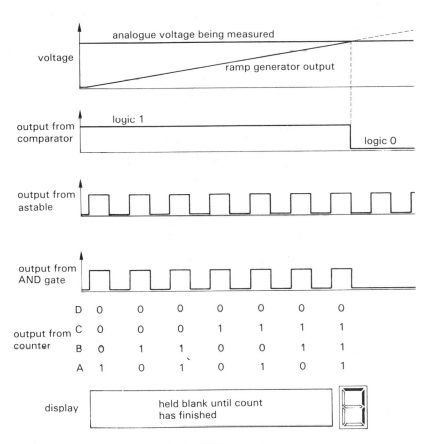

The voltage to be measured is put on the input.
The ramp generator starts to produce a rising voltage.

The comparator has a high output until the ramp reaches the same voltage as the input.

While this has been going on, the astable has been producing square waves.

The AND gate output will follow this until the comparator produces logic 0.

The counter will stop counting pulses when the AND gate output goes to 0.

The display shows the number counted to.

Figure 27.2 Timing diagram for the DVM

If a larger voltage is put on the input, the ramp generator will take longer to reach it and the number counted to will be larger.

Although this seems to be quite a long process, the frequency of the astable can be very high. A DVM typically takes about 1 second to do the A to D conversion and produce a reading. There are other types of A to D converters which work even faster.

Other uses of A to D conversion

The digital signal from an A to D converter could be used as an input to a computer. This allows the computer (which works on digital signals) to be **interfaced** (or connected) to measuring devices and sensors which give an analogue voltage.

28
Exercises

1 Signals often need to be transmitted across a distance.
(a) One way of doing this is shown in Figure 28.1:

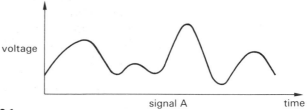

Figure 28.1

(i) Is signal A an 'analogue' or a 'digital' signal? (ii) Give a reason for your answer.
(b) Another type of signal is shown in Figure 28.2:

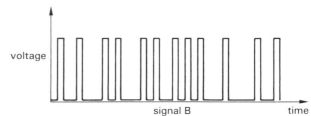

Figure 28.2

(i) What type of signal is this? (ii) Give a reason for your answer.
(c) We can turn signal A into the type of signal B, transmit it in that form and then turn it back into its original form. One way of doing this is shown in Figure 28.3:

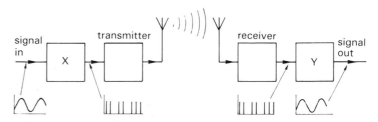

Figure 28.3

(i) What is the name of the circuit labelled 'X'? (ii) What is the name of the circuit labelled 'Y'? (iii) Explain carefully why this method of transmission is more reliable than the method shown in Figure 28.1.

2 Explain, with examples, what is meant by 'binary' and 'decimal' numbers. Why are digital systems particularly useful for calculations using binary numbers?

3 What are the main differences between LCD and LED displays?

4 Explain, with a diagram, what is generally meant by 'logic level 1' and 'logic level 0'.

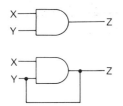

Figure 28.4

5 (a) Draw the truth table for the two-input AND gate in Figure 28.4
(b) If the output of an AND gate is joined to one of its inputs its behaviour will be changed. (i) What will be its output if both inputs are connected to logic 1? (ii) What will be its output if input X stays at logic 1 but the logic input is disconnected from input Y?
(c) The circuit in Figure 28.5 uses three AND gates connected as in Figure 28.4. The relay is energised when three inputs, P, Q, and R are each connected for a moment, in sequence, to logic 1. Calculate the voltage at point Z (assume the inputs to the AND gates take no current).

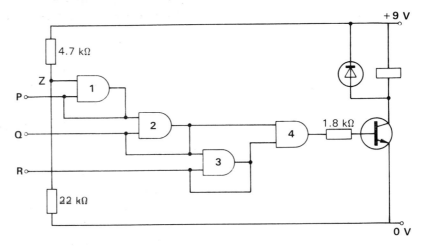

Figure 28.5

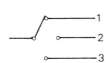

Figure 28.6

(d) Initially the output of each AND gate is at logic 0. Explain clearly, by using a truth table or otherwise, what happens to the inputs and outputs of AND 1, AND 2 and AND 3 when: (i) first input P is connected to logic 1 and then disconnected; (ii) input Q is next connected to logic 1 and then disconnected; (iii) input R is next connected to logic 1 and then disconnected.
(e) Explain why, at the end of the sequence described in part (d), the relay coil is energised.
(f) This circuit could be used as a three-digit combination lock because the relay coil is energised only when the correct sequence of switching is applied. Suggest a switch circuit, based on a rotary switch, as shown in Figure 28.6, and any other necessary components, that would operate the relay coil when the number 231 was inserted.
(g) Show, with the aid of a diagram of the appropriate parts of the circuit, how you would add to or alter the circuit to (i) switch off the relay (ii) reset all the AND gate outputs to logic 0. (*AEB 84 paper 2*)

Section 4

Audio Systems

29
Types of Signals

A **signal** is a way of sending **information**. There are many obvious examples such as railway signals, smoke signals or simply waving goodbye to someone! To send information, something must **change**. For example a hand is lifted up or a red light goes on.

Electronics and information

One of the most important uses of electronics is to take information, perhaps store it or transmit it, and finally get the information back again. Examples of systems which do this are record players, radios, computers, TVs and so on.

This **information** might be sound waves, printing, light, etc., but inside electronic circuits the information must be transmitted in the form of **electrical signals**. This is usually done by changing the voltage at a point.

Waveforms

A **waveform** is a diagram which shows how the signal changes as time goes by. It is a graph of voltage plotted against time. The size of the change is called the **amplitude** (*A*) of the signal. Some common waveforms and their uses are shown in Figure 29.1.

The ground marshall communicates with the pilot by waving two large bats. In many cases simplest signals are the most effective. They are also less likely to be misunderstood.

TV aerials produce electrical signals from radio waves.

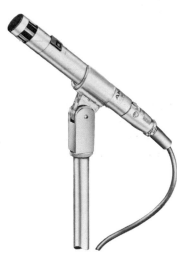

A microphone is a common source of electrical signals.

SHAPE	TYPICAL SOURCE	USE
sine wave	1. dynamo 2. sine wave oscillator	a.c. mains
square wave	astable	lamp flasher
triangular wave	integrator with square wave input	electronic music
sawtooth wave	sawtooth oscillator	oscilloscope timebase
ramp	integrator with pulse input	analogue-to-digital conversion
positive going pulse negative going pulse	switch	triggering latches and monostable
pulse train	telephone dial	data transmission

Figure 29.1 Some common waveforms and their uses

Signal sources

A device which produces a signal is called a **signal source**. Examples include microphones, astables, TV aerials, record pick-ups and so on.

Summary

● Signals carry information.
● An electrical signal can be a change in voltage or current.
● The size of the change is called the amplitude of the signal.
● Signal sources are devices which produce signals.

Questions

1 Look at the sine wave in Figure 29.2.
    (a) What is the amplitude?
    (b) How many complete cycles are shown?
    (c) How long does each cycle take?
    (d) What is the frequency of this sine wave?
    (e) Draw a sine wave with twice the frequency.

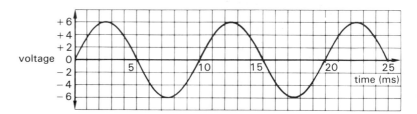

Figure 29.2

2 Which of the waveforms in Figure 29.1 would be best for each of the following?
    (a) a digital clock
    (b) the input to a monostable
    (c) a metronome
    (d) the input to a robot arm to make it wave goodbye.

30
Audio Systems

Audio signals

A **sound wave** is a pattern of air vibrations. An **audio signal** is an electrical copy of these vibrations: a changing voltage instead of a movement of air. The audio signal which corresponds to the sound of a flute is a simple sine wave, but other sounds have much more complicated patterns. See Figure 30.1.

The audio signal produced by the London Philharmonic Orchestra is a mixture of signals from many different instruments. An audio system must be able to process a wide range of complex signals accurately.

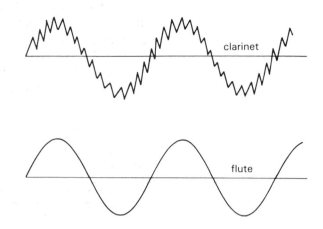

Figure 30.1   Typical waveforms from two instruments playing the same note

Audio systems

Once a sound wave has been copied into an electrical audio signal, many things are possible. The signal can be increased (or **amplified**), it can be stored (perhaps by using it to make magnetic patterns on a magnetic tape) or it can be modified and eventually turned back into a quite different sound. All these processes are carried out in **audio systems**.

Block diagram of an audio system *(Figure 30.2)*

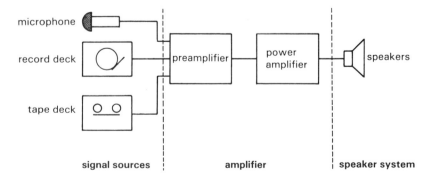

Figure 30.2   Block diagram of an audio system

We can look at each part of this system in turn.

The Lansdowne Recording Studios in West London. The mixer shown here can control up to 40 different input signals at once to produce just two output signals. What other parts of an audio system can you see in this photograph?

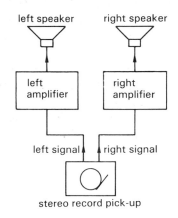

Figure 30.3 Block diagram of a stereo record player system

Signal sources

The microphone, record player and tape deck are all signal sources. They produce audio signals; either directly from sound waves (in the case of the microphone) or from recorded sound.

Amplifier section

The **preamplifier** increases the amplitude of the audio signal from the signal source. It will also probably contain circuits to adjust the 'tone' (and 'balance' if it is a stereo system).

The **power amplifier** or **output amplifier** amplifies the signal further and provides a large current to drive the speaker system.

Speaker system

This contains one or more loudspeakers (and perhaps a circuit called a **crossover**). The speakers convert the electrical audio signals back into sound waves.

Signal mixer

This is used to combine signals from more than one source. Signal mixers are used in recording studios or on stage where signals from several microphones are mixed together.

Mono and stereo

A **mono** system has only one signal passing through it. Although it may use more than one speaker, the sound from each one is the same. A **stereo** system has two completely separate signals passing through. This means that apart from a stereo signal source, two amplifiers and two sets of speakers are needed. A different sound comes out of each speaker. (See Figure 30.3.)

Summary

● An audio signal is an electrical copy of a sound wave.
● An audio system processes audio signals and produces sound.
● A mono system produces only one signal.
● A stereo system produces two different signals.
● A signal source produces the audio signal.
● This is amplified before being turned into sound by speakers.

Questions

1 Draw a block diagram of an audio system, explaining what each part does.
2 What is the difference between mono and stereo systems?
3 What would you need to add to a mono system to turn it into a stereo system?
4 Describe a use of an audio system (apart from listening to music).

31
Microphones

Microphones are transducers which produce electrical audio signals from sound waves. There are two main types of microphones:

1 **Active microphones** produce a changing e.m.f. which can be used as a (voltage) signal.
2 **Passive microphones** do not produce their own e.m.f. Instead they need a steady power supply and then produce a changing voltage or current from it.

Active types
Moving-coil (*Figure 31.1*)
Sound waves striking the diaphragm make it (and the attached coil) vibrate. The movement of the coil in the field of the magnet induces a changing e.m.f. in the coil at the same frequency.

Crystal microphone (*Figure 31.2*)
Crystal microphones use the **piezoelectric effect**. When a special type of crystal sandwich (or **bimorph**) bends it produces a small e.m.f. Vibrations of the diaphragm are transferred to this special crystal which also vibrates, producing a signal.

Differences between types of active microphone
● Moving-coil microphones are most commonly used for recording and on stage. They can be either 'high impedance' (47 kΩ) or 'low impedance' (600 Ω).
● Crystal microphones are very cheap and can be made very small. They have a very high impedance (1 MΩ) and give a large signal. They have a poor frequency response.
● 'Ribbon' microphones are still found in some recording studios. They are the most sensitive. They should not be 'blown' into!

A typical moving-coil microphone.

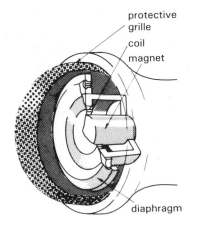

Figure 31.1 Construction of a moving-coil microphone

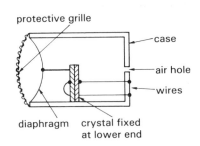

Figure 31.2 Construction of a crystal microphone

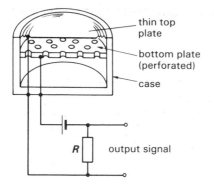

Figure 31.3 Construction of a condenser microphone

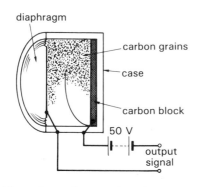

Figure 31.4 Construction of a carbon microphone

Passive types

Condenser (capacitor) type (*Figure 31.3*)

The two plates behave like a capacitor. A small battery is connected across them. Vibrations of the top plate (which acts as the diaphragm) alter the distance between the two plates and therefore alters the value of the 'capacitor' which charges and discharges through the resistor. The changing p.d. across the resistor is amplified before being sent out as the signal.

The built-in microphone on portable cassette recorders is usually this type.

Carbon (*Figure 31.4*)

Vibrations of the diaphragm make the moving carbon block compress and release the carbon grains. This affects the resistance between the two blocks which alters the current flowing (from a 50 V d.c. supply) through the circuit and creates a signal.

Carbon microphones give a large, poor-quality signal and are only used in the telephone system where a 50 V supply is available.

Directional properties of microphones

Unidirectional microphones are much more sensitive to sound coming from a particular direction. This is useful for picking out particular sounds from other nearby sources.

Omnidirectional microphones pick up sound from a wide angle.

Questions

1 Which types of microphone produce their own e.m.f?
2 Draw diagrams and explain how two different types of microphone work.
3 Which microphone has a very high impedance?
4 Why cannot microphones be connected directly to a speaker?

Recording a TV programme. Why do you think that a unidirectional microphone is used here?

An omnidirectional microphone will make sure that the signal does not alter if the performer moves from side to side.

32
Loudspeakers

A loudspeaker is a transducer which turns electrical audio signals into sound waves.

How a speaker works

Look at Figure 32.1. If an alternating current (for example an audio signal) flows in the coil, the coil vibrates inside the fixed magnet.

The **cone** attached to the coil also vibrates. This produces sound waves at the same frequency as the audio signal. Bigger signals will cause larger vibrations and louder sounds. Note that a steady direct current will not produce any sound (apart from a click when it first starts flowing).

Speakers need quite a large current to work and the amplifier must be able to provide this.

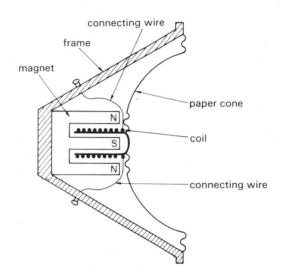

Figure 32.1 Cross-section through a moving-coil loudspeaker

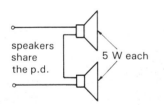

10 W total in each case

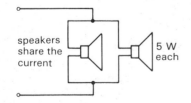

Figure 32.2 Two speakers sharing the load can together handle twice the power of a single speaker

Power rating

The **power rating** of a speaker is the maximum amount of electrical power that it can handle before overloading. Typical power ratings are 5 W, 10 W, 25 W, etc. It is better to use a speaker with a power rating larger than the amplifier's output power. Two or more speakers together will give a larger overall power rating. Each speaker takes a share of the total load. See Figure 32.2.

Impedance

The **impedance** (or resistance) of the loudspeaker's coil must **match** the amplifier it is being used with. (See chapter 35.) Suppose that an amplifier needs an 8 Ω loudspeaker. When connecting speakers together the overall impedance must still 'match' the amplifier. If you are not sure, the safest way to avoid damage to the amplifier is to connect speakers in series. See Figure 32.3.

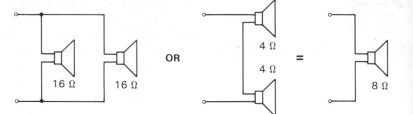

Figure 32.3 Two 16 ohm speakers in parallel or two 4 ohm speakers in series behave like a single 8 ohm speaker

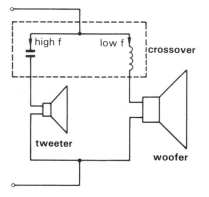

Figure 32.4 Woofer and tweeter connected in a simple crossover system

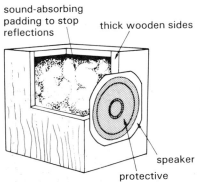

Figure 32.5 Construction of a typical loudspeaker cabinet

Frequency response

The **frequency response** of the speaker describes the range of signal frequencies which the speaker can reproduce properly. It depends on a number of things like the strength of the magnet and the diameter and mass of the cone.

Woofers are large speakers with heavy cones. They are best for low frequencies (say 40 Hz to 500 Hz).

Tweeters are small speakers with light cones which can move faster. They are best for high frequencies (say 500 Hz to 5 kHz).

One way to obtain a wide frequency response is to use a woofer and a tweeter together in a **speaker system**. Look at Figure 32.4. The circuit shown inside the dotted lines is called a **crossover**. Low frequencies cannot pass through the capacitor. High frequencies cannot pass through the inductor. The overall signal is **filtered** so that each speaker receives the correct frequency signals only.

Cabinet design

Speakers should be mounted in a box or cabinet. (See Figure 32.5.) As well as moving air, the speaker also makes the sides of the cabinet vibrate. This increases the total amount of sound. The effect is called **resonance**. A good cabinet is made of a rigid material like **chipboard**. A material like fibreglass wool can be fixed inside the cabinet to absorb unwanted reflections.

Summary

● A loudspeaker uses the electromagnetic effect to turn electrical signals into sound.
● The amount of sound produced depends on the output power of the amplifier, the size of the speaker and the cabinet used.
● For a good frequency response, a woofer and a tweeter can be connected with a crossover to make a speaker system.
● The impedance of the speaker system must match the amplifier.

Questions

1 Draw a diagram and explain how a speaker produces sound.
2 Draw a diagram and explain why it is important to match the impedance of the speaker system to the amplifier.
3 Draw a diagram of a woofer and tweeter connected by a crossover. Explain why this is done and how it works.
4 An elliptical (oval) speaker can give a reasonable frequency response without taking up much space. Where are they used?

33
Disc Recording

A 'gramophone' record is a way of storing sound by cutting a wiggly groove in a disc of plastic. When a sharp point (or **stylus**) moves along this groove it vibrates in the same way that the air was vibrating when the original sound was made.

In the original 'phonographs' all you heard was the stylus vibrating (amplified by a metal horn), but in a modern record player these vibrations are converted into electrical audio signals and processed by an audio system. (Although you can still hear the stylus vibrating if you listen carefully enough.)

How a disc is recorded *(Figure 33.1)*

signal makes stylus vibrate

'wiggly' grooves

signal on master tape

section through the master disc

acetate disc

stylus on cutting head

die

master

copy

die

die

die

copies are made by stamping with a die made from the master

Figure 33.1 Simplified diagram of the recording process

Making the original 'acetate' on a disc-cutter lathe. The air and equipment must be kept clean; even small dust particles may damage the soft vinyl disc and cause noise on the finished record.

How a disc is played *(Figure 33.2)*

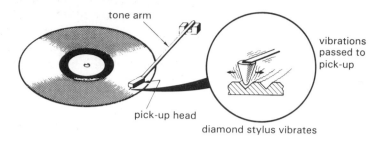

Figure 33.2   Simplified diagram of the playback process

The pick-up is a transducer which produces electrical signals. Vibrations of the stylus are passed to a small magnet and coil.

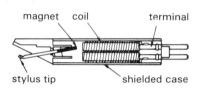

Figure 33.3   A magnetic pick-up

Magnetic pick-ups
In some cases the coil is fixed and the magnet vibrates. (See Figure 33.3.) Other pick-ups keep the magnet fixed and let the coil move. Either way, the vibrations induce a signal (about 5 mV) in the coil. Many magnetic pick-ups are in a **cartridge** which can be easily changed for another.

A moving-magnet pick-up inside its cartridge. Notice the four connecting pins at the back. Is this mono or stereo?

Ceramic pick-ups
These use the vibrations of a special crystal to produce the signal. Although the signal is large (typically 300 mV), the frequency response is not very good. Although ceramic pick-ups are not used much now, some hi-fi systems still allow either pick-up. A switch selects which preamplifier to use.

Equalisation
Because of the way a magnetic pick-up operates it does not produce the same size signal at all frequencies. If its signal was amplified directly there would be more treble than bass. The signal must be equalised by a circuit in the preamplifier to make it sound normal. Ceramic pick-ups behave in a different way and do not need the same equalisation.

Stereo
Most records are **stereo** and have two separate signals. One is recorded on each side of the groove. A stereo pick-up has two magnets or coils at right-angles to each other. One reacts to the left side and the other to the right side. Two separate signals are produced, one for each **channel** of sound.

Summary
- A disc is recorded by cutting a wiggly groove which is a copy of the way the air vibrated during the original sound.
- When the record is played a stylus is made to vibrate.
- Vibrations of the stylus are converted into audio signals.
- Stereo records have a set of wiggles on each side of the groove. They produce two different audio signals.

Questions
1 Look at the picture of the back of an amplifier in Figure 33.4.
   (a) What do 'cer. P/U' and 'mag. P/U' stand for?
   (b) Why are there two sockets for each one?
   (c) Why does the amplifier treat the signals from the two types of pick-up differently?
   (d) How would it sound if a magnetic pick-up was used with a preamplifier designed for ceramic pick-ups?
2 What is the difference between mono and stereo?
3 Describe what things can damage a disc and how they can be avoided.

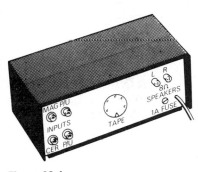

Figure 33.4

34
Tape Recording

Tape recording consists of converting electrical signals into a pattern of magnetism on a magnetic tape. Playback consists of turning these patterns of magnetism back into electrical signals.

Magnetic tape

This is a plastic tape coated with extremely small magnetic particles (iron oxide or chromium dioxide). Before use the magnetic poles of the particles point in all directions on the tape. See Figure 34.1.

Recording

The electrical signal to be recorded is sent into the coil in the **tape head** (Figure 34.2) and produces a magnetic field around the thin gap in the head. As the tape passes in front of the gap, the poles of the particles are arranged into a set pattern by this field.

The particles will point left or right, depending on which way the current was flowing in the coil when they passed. See Figure 34.3.

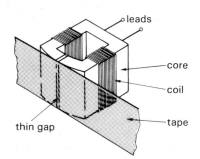

Figure 34.1 Microscope photograph of a piece of Sony HF tape. The particles are so small that ten thousand of them side-by-side will still measure less than a millimetre wide.

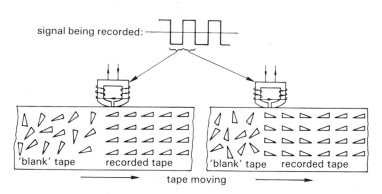

Figure 34.3 Diagrammatic view of the tape before and after recording a signal. In reality the pattern is too small to see.

Figure 34.2 The tape head. Notice the gap which is extremely thin (typically about 0.001 mm). The surface is highly polished to reduce wear on the tape.

Playback

A switch connects the coil of the head to the input of an amplifier. The recorded tape passes by the tape head causing a changing magnetic field as each set of magnetic particles passes. This induces an alternating current in the coil. If the speed of the tape is the same as before (4.75 cm/s in cassette recorders), the frequency of this signal will be the same as the signal which made the original recording. See Figure 34.4.

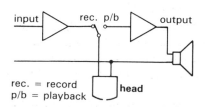

Figure 34.4 Block diagram of the tape recorder

track arrangement for mono recording on both sides

(a)

track arrangement for stereo recording on both sides

(b)

Figure 34.5 Recording on both sides and in stereo

low amplitude signal—
not all particles are
lined up

high amplitude signal—
all particles lined up

high frequency signal—
makes a closer pattern

Figure 34.7 The effect of amplitude and frequency on the recorded pattern on the tape

Recording on both sides and in stereo

The tape head is only half the width of the tape and records on the bottom part of the tape only. When the tape is turned over, the other half passes in front of the head. (Figure 34.5a)

A stereo tape head has two coils and cores; one for each channel (or **track**). Each track takes up half the width of a 'side' (or a quarter of the whole tape width). (Figure 34.5b)

Multitrack recorders

In recording studios tape recorders with up to 32 tracks are used. Each track is used for a different instrument. These are then **mixed down** into a stereo recording for the final record. Figure 34.6 shows the head of a multitrack recorder.

Figure 34.6 The head of a multitrack recorder which can record sixteen separate tracks at once

Erasing

A separate **erase head** is mounted just before the record/playback head. When the machine is recording, a high-frequency signal (above the range of hearing) is sent into the erase head. This destroys any existing patterns so that a new signal can be recorded. This is switched off during playback.

Limits to amplitude and frequency

Figure 34.7 shows that the **amplitude** of the recording signal affects the **number** of particles that are lined up. The **frequency** of the signal determines **how close together** the pattern is. There is a limit to the amplitudes and frequencies that can be faithfully recorded.

At a certain amplitude of the recording signal, **all** the particles are lined up. The tape is said to be **saturated**. Any greater signal than this will simply be distorted. To avoid this, the recording signal must always be kept below this level.

0.04 mm

each pair of tracks contains the information needed to produce one picture

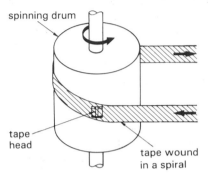

spinning drum

tape head

tape wound in a spiral

Only one head is shown. The other is on the opposite side of the drum.

Figure 34.8 Simplified diagram of the VHS principle

The highest frequency that can be recorded depends on how fine the pattern can get. This depends on the size of the particles and how fast the tape moves. Early tape recorders moved the tape along at high tape speed to spread the pattern out but now improvements in magnetic tape using much smaller particles allow much slower tape speeds to be used. The limit for audio cassette tapes is currently about 10 kHz.

Video recording

This works on the same principle but the signals being recorded come from a video camera (or TV tuner). Each TV picture contains 625 **lines** and 25 pictures are produced each second. Very high frequencies are needed to record all this information!

This used to be done by using a very high tape speed but now we use a system called video helical scan (or VHS). Each line is recorded as a diagonal stripe on the tape, produced by making the tape head move in a spiral (or **helix**). (See Figure 34.8.)

Uses of tape recording

● Recording for entertainment; not only in the home but in recording studios and so on.
● Recording signals from unattended scientific measuring instruments for later analysis.
● A storage system for data in some home computers or the control of automatic robots in some factories.

Summary

● Tape recording consists of converting signals into magnetic patterns on a moving tape using a tape head.
● The signals can be recovered by playing back the tape. The magnetic pattern induces a signal in the tape head.
● There is a limit to the highest frequency that can be recorded. This depends on the type of tape and its speed.
● Stereo recordings are made on two tracks. Each track takes up part of the width of the tape.
● Video recorders overcome the problems of very high-frequency signals by recording in diagonal stripes along the tape.

Questions

1 Draw a diagram and explain what magnetic tape is made from.
2 When we record on a tape, what happens to the magnetic particles?
3 Why is there a limit to the highest frequency that can be recorded on a tape?
4 Explain, using diagrams, how we can record on both sides of a tape.
5 Why does a stereo-recorded tape still work on a mono tape recorder?

35
Matching

When we connect up systems using signal sources and so on we must make sure that the output of one circuit can feed the input of the next one properly. This is called **matching**.

Connecting an 'ideal' signal source to a load

A signal source produces an e.m.f. which drives current through the load. With an ideal signal source all the e.m.f. it produces ends up as a p.d. across the load. (See Figure 35.1.)

'Real' signal sources

Every signal source has some **impedance**. This comes from the resistance of its wires and components. (We use the word **impedance** rather than **resistance** when considering a.c. signals.) When a current flows into the load some p.d. will be 'lost' inside the signal source. The p.d. across the load will be less (Figure 35.2). If more current flows then even more p.d. will be lost.

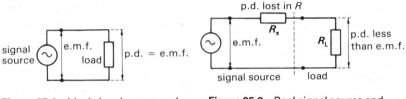

Figure 35.1   Ideal signal source and load

Figure 35.2   Real signal source and load

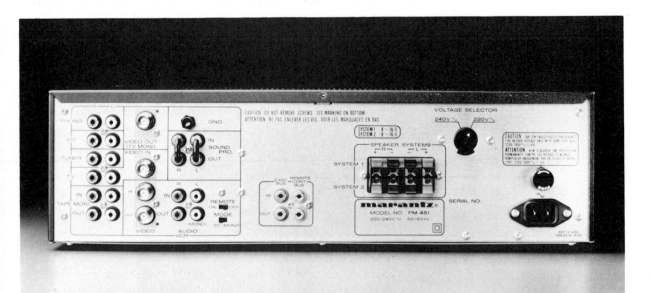

A good hi-fi amplifier has many different input circuits. Each one is designed to match a particular signal source. Matching is important at the output as well. What speakers are recommended for this amplifier?

Connecting a signal source to a load

There are several ways to make sure that the maximum amount of signal ends up in the load and is not wasted inside the signal source:

● We get the maximum amount of p.d. across the load if the load has a very **high** impedance (Figure 35.3).

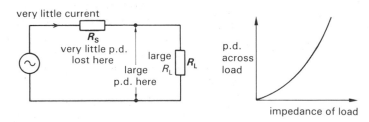
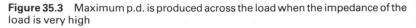

Figure 35.3   Maximum p.d. is produced across the load when the impedance of the load is very high

● We get the maximum amount of **current** flowing in the load if the load has a very **low** impedance (Figure 35.4).

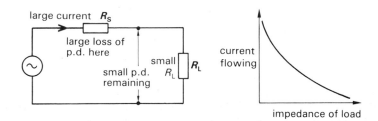
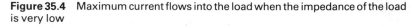

Figure 35.4   Maximum current flows into the load when the impedance of the load is very low

● We get the maximum amount of **power** produced in the load if the load has the **same** impedance as the signal source (Figure 35.5).

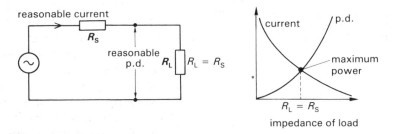
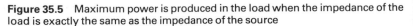

Figure 35.5   Maximum power is produced in the load when the impedance of the load is exactly the same as the impedance of the source

36
Noise

Noise is an unwanted signal produced inside a system. It interferes with the signals we want and can sometimes completely swamp them. All systems produce some noise. The **signal-to-noise ratio** of a system describes the **amplitude of the wanted signal** compared with the **amplitude of the noise** present.

We shall now look at different types of noise and their causes.

Random noise

This includes the hiss heard when tuning-in a radio between stations or when listening to quiet passages in music, 'snow' on a television picture, and the ragged edges to waves seen on an oscilloscope when examining very small signals.

Causes of random noise

● **Electrical component noise** is generated inside the actual components. It is caused by tiny random electric currents, mainly inside the resistors. The effect increases with temperature. the answer is to use good-quality, **low noise** components. For example **metal oxide** resistors are better than **carbon** types.

● Scratched and dusty surfaces of gramophone records will produce **surface noise**. The answer is to keep the surface clean and free of grease which will make dust stick.

● Small movements of the magnetic particles on recording tape will produce **tape hiss**. The answer is to use good-quality tape and keep the signals as high as possible (without overloading).

Unwanted signal pick-up

The hum or buzz sometimes heard from speakers, clicks and pops when other equipment is operated, sudden breakthrough of radio transmissions and so on are all examples of unwanted signals being picked up by the system from outside.

Causes of unwanted signal pick-up

● Wires in the circuits, connecting leads and even people can act as aerials, picking up unwanted r.f. (radio) signals. The answer is to put all circuits inside a metal case and connect this case to earth. Special **screened** cables (see chapter 78) should be used for carrying small signals, especially at the input.

● A humming or buzzing sound is caused by the circuit picking up the electromagnetic waves produced by transformers and mains leads. It can also be caused by insufficient **smoothing** when turning the a.c. mains into the low-voltage d.c. for the power supply (see chapter 51). It can be reduced by mounting transformers well away from the rest of the circuit. Special 'toroidal' transformers are available which give out fewer waves. An earthed metal case also reduces mains hum.

● Interference from other mains equipment (e.g. a washing-machine) can be reduced by putting **filters** in the mains leads and **suppressors** in the other equipment.

Although there is a faint picture on this screen, the signal has been almost completely swamped by noise.

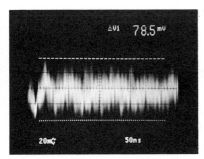

Even a steady d.c. from a power supply contains some noise. This picture was taken from an oscilloscope adjusted to pick out very high-frequency signals with a small amplitude.

An example of amplifier design

Figure 36.1 is a view inside a high-quality amplifier. Notice how the manufacturers have tried to reduce the noise.

The metal case is an important screen against noise. A good screen covers as much of the circuits as possible.

Figure 36.1 Inside a high-quality amplifier. Notice the position of the power supply transformer and the connecting wires. How else have the manufacturers tried to reduce unwanted noise?

Questions

1 What would you hear if an audio system suffered from:
   (a) component noise
   (b) r.f. pick-up
   (c) mains hum?
2 Does a good system have a large or small signal-to-noise ratio?
3 What kinds of noise can be reduced by:
   (a) using an earthed metal case
   (b) keeping the signal large
   (c) using metal oxide resistors?

37
Electronic Sounds

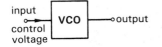

Figure 37.1 Voltage-controlled oscillator (VCO)

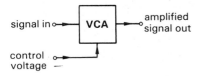

Figure 37.2 Voltage-controlled amplifier (VCA)

It would be impossible to produce most of the sounds used in popular music and films today without electronics. In this chapter we will see how some of the special effects are created.

'Echo'

Echoes can be produced by using a special type of tape recorder. The signal is recorded and then played back straightaway. Several playback heads may be used to give more than one echo.

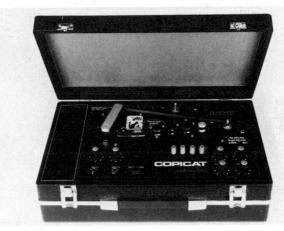

This tape-loop echo chamber has three playback heads. The Watkins' 'Copicat' was responsible for many of the distinctive sounds of popular music during the '60s and '70s although now it has largely been replaced by digital effects.

Electronic noises

Most sounds can be broken down into simple, basic noises. These can be produced by the building blocks described below. If we combine these into systems, we can **synthesise** sounds.

- **Voltage-controlled oscillator (VCO)**. The **frequency** of the signal produced can be varied by changing the **voltage** on the input. (Figure 37.1)
- **Voltage-controlled amplifier (VCA)**. The **amplitude** of the output signal can be varied by a **control voltage**. (Figure 37.2)
- **Waveform generator**. This produces a variety of different signals, e.g. sine, square or triangle wave signals. (Figure 37.3)
- **Noise generator**. This produces a random **white noise** (like a 'hiss'). (Figure 37.4)

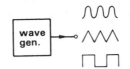

Figure 37.3 Waveform generator

Figure 37.4 Noise generator

We shall now look at some examples of how sounds can be built up using these circuits.

A fully electronic synthesised drum kit. The 'drums' can be any shape. They are simply sensors which send a signal to the synthesiser when struck. Apart from producing many different sounds, the kit is much more portable than conventional drums.

Siren

The waveform generator alters the frequency of the VCO to give a rising and falling note (Figure 37.5):

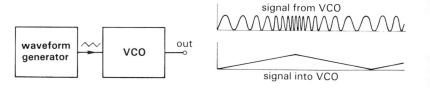

Figure 37.5 Block diagram of how to obtain a 'siren' sound

Steam train (chuff chuff!)

The waveform generator turns the VCA on and off, which pulses the signal from the noise generator (Figure 37.6):

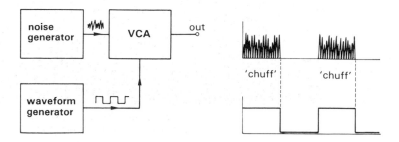

Figure 37.6 Block diagram of how to obtain a 'steam train' sound

Musical sounds (e.g. piano)

When a note is played on a piano, the sound starts suddenly and then gradually dies away. See if you can work out how this block diagram can produce this type of signal (Figure 37.7):

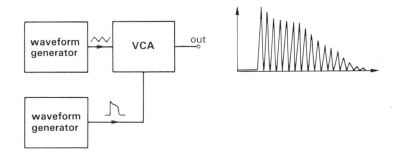

Figure 37.7 Block diagram of how to produce a piano sound

Synthesiser

A **synthesiser** contains all of the building blocks described (and many others). Try to identify some of them in the photograph.

A typical synthesiser controlled by a keyboard. The VCO is called a DCO (digitally-controlled oscillator) on this model. What other sections can you recognise?

38
Digital Audio Systems

All of the audio systems described so far have been analogue and have used analogue signals. Even with the best equipment, some unwanted noise is bound to creep in as the signals pass through the system. Because digital signals do not suffer from this problem many audio systems now turn the analogue audio signals into digital signals for at least part of their journey through the system. This generally needs two types of circuit:

Analogue-to-digital (or A to D) converter

The changing voltage level of the analogue signal is 'sampled' and measured thousands of times each second. Each time, the value is turned into a binary number. The number is sent out as a digital signal. (Figure 38.1)

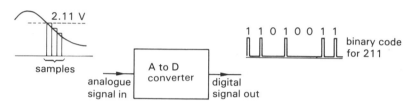

Figure 38.1   Action of an A to D converter

Digital-to-analogue (or D to A) converter

The D to A converter reverses the process. The voltage level at the output can vary between 0 V and +5 V. Its actual value depends on the digital signal sent into the input. A stream of different digital signals produces a changing analogue signal at the output. (Figure 38.2)

Digital recording

The analogue signals from microphones, etc. are turned into digital codes by A to D converters.

The digital signals can be mixed together and then recorded onto tape or used to make a compact disc recording. Not only do the signals remain free of noise and distortion but also, being digital, they can be controlled by a computer. This means that very complicated mixing and special effects can be made easily.

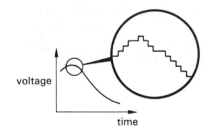

Figure 38.2   The output from a D to A converter is really a series of small steps in voltage

Compact discs (or laser discs)

These do not have a groove. The signal is recorded by making very small reflective patterns on a plastic disc.

Light from a laser beam is shone onto the spinning disc, and each time that it strikes one of the shiny patches it is reflected back to a sensor. This produces a stream of digital pulses, one pulse per reflection. These are sent through a D to A converter and processed by a normal audio system.

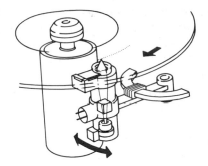

Figure 38.3 shows how a compact disc player operates.
The signal quality from a compact disc is very good because:
● There is no stylus to damage the record or wear it out.
● It does not matter if the surface gets scratched or dirty.

Figure 38.3 The spinning compact disc is read by a laser underneath which is gradually moved across the disc by a servo motor

Digital echo *(Figure 38.4)*
The signal is split into two. Part of it is converted to a digital signal. This signal is then passed along a long chain of circuits. This takes some time. When the digital signal is converted back it will be slightly 'behind' the original signal and this gives the echo effect.

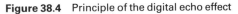

Figure 38.4   Principle of the digital echo effect

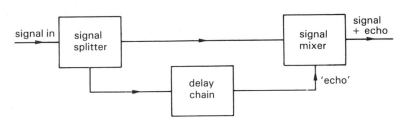

A digital echo chamber.

A ROM unit which plugs into a Casio keyboard.

Digital synthesisers
These can produce digital audio signals directly. Some create the signals from computer programs. Others can store sounds as digital signals in a computer memory and output them when needed, perhaps in a completely different key to the original sound.

ROM storage
Complete pieces of music can be stored as digital signals in a **read-only memory** (or ROM). See chapter 66. Although a huge amount of data needs to be stored for even quite short recordings, technology is steadily creating more memory space on smaller silicon chips.

39
Exercises

1   Shown in Figure 39.1 is a trace from an oscilloscope screen.
The timebase was set at 10 ms/cm.
The Y-gain was set at 5 V/cm.

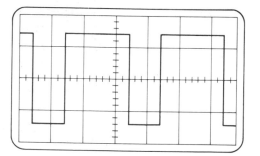

Figure 39.1

(a) What is this type of waveform called?
(b) Name one circuit or device that can produce this waveform.
(c) State one use of this type of waveform.
(d) What is the amplitude of the wave shown?
(e) How long does one cycle take?
(f) What is the frequency of the signal?

2   Figure 39.2 shows a block diagram of a record player:

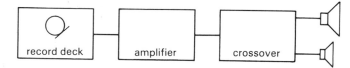

Figure 39.2

(a) Is the system mono or stereo?
(b) Why are there two speakers of different sizes?
(c) What does the crossover do?
(d) Describe briefly how a magnetic or a ceramic pick-up produces a signal when it is being used to play a record.

3   (a) Describe briefly how signals are stored on magnetic tape.
(b) Describe three types of information or signals that can be stored in this way.

4   (a) Name three signal sources that use the electromagnetic effect.
(b) Name two signal sources that use the piezoelectric effect.

5 The block diagram in Figure 39.3 shows a signal source connected to an amplifier which is connected to a speaker.
The input impedance of the amplifier is 48 kΩ.
The output impedance of the amplifier is 16 Ω.

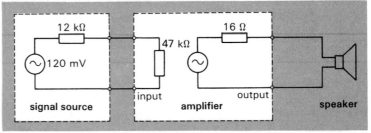

Figure 39.3

(a) Looking at the **input** to the amplifier. (i) Calculate the total resistance met by the signal from the signal source. (ii) Calculate the current flowing into the input of the amplifier. (iii) Calculate the voltage of the signal at the input of the amplifier. (iv) What changes would increase the voltage of the signal arriving at the amplifier (apart from an increase in the signal produced by the signal source)?
(b) Shown in Figure 39.4 are some combinations of speakers. Each speaker is rated at 5 W.

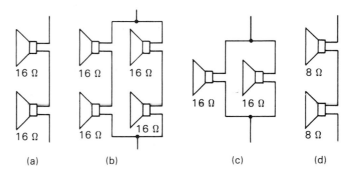

Figure 39.4

(i) What is the impedance of each of the combinations shown?
(ii) Which combinations will best match the amplifier in part (a)? (iii) Which combinations may cause damage to the amplifier? (iv) What is the advantage of combination (b) over combination (d)?

6 'Noise' can be a problem in many systems.
(a) Describe three ways in which noise can arise in a system.
(b) Describe the effect of noise in each of the following: (i) A tape recorder being used to record from a microphone. (ii) An oscilloscope being used to look at a small signal. (iii) An audio system playing a gramophone record.
(c) For each of the above, state any ways that the signal-to-noise ratio can be improved.

Section 5

Analogue Systems

40
Amplification

We often need to make a small signal larger. We can do this with an **amplifier**. A small input signal is sent into the input. This signal is used to control a larger amplified signal which appears at the output. The output signal draws its energy from the power supply. See Figure 40.1.

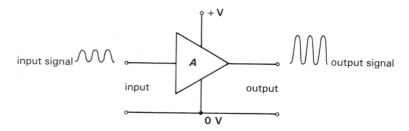

Figure 40.1 General behaviour of an amplifier

Gain

The amount of amplification produced is called the **gain** of the amplifier. Gain is usually abbreviated to 'A' and is calculated by:

$$\text{Gain } (A) = \frac{\text{size of output signal}}{\text{size of input signal}}$$

A typical calculation is shown in Figure 40.2.

It is important to realise that we are not simply making a small signal larger (this would be magic!), but we are making a (larger) **copy** of the input signal. We make this copy from the steady voltage of the power supply using the input signal as a 'pattern'.

If output signal $= 100\,mV$
when input signal $= 5\,mV$
Gain $(A) = \dfrac{100}{5}$
$= 20$

Figure 40.2

Negative gain *(Figure 40.3)*

In some types of amplifier the output signal is **out of phase** with the input signal. The output waveform is 'upside-down'. We describe this by writing the gain as a negative value. Note that this does not mean that the output is smaller than the input!

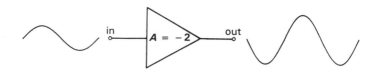

Figure 40.3 Behaviour of an amplifier with negative gain

Cascading *(Figure 40.4)*

Two or more amplifiers can be connected together so that the output of the first stage is sent into the input of the second stage and so on. This is called **cascading**. It gives more gain than can be obtained from a single amplifier.

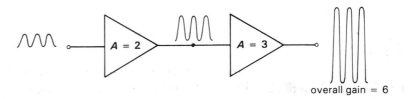

Figure 40.4   Cascading two amplifiers

Clipping *(Figure 40.5)*

The e.m.f. and current available from the power supply sets a limit on the size of the output signal. If the output signal tries to go above this limit a problem called **clipping** arises. It occurs when either the gain or the input signal is too large.

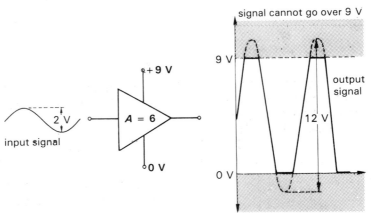

Figure 40.5   Clipping in an amplifier

Effects of clipping

A highly clipped signal becomes a square wave. Sometimes this is used as a way of producing sharp square waves but usually it is to be avoided. If the signal was a piece of music for example the result of clipping is a harsh, distorted sound.

Clipping can be prevented by keeping the gain down to a sensible level and/or reducing the size of the input signal.

Uses of amplifiers

The photographs on this page show some of the many uses of amplifiers.

The computerised cardio-diagnostic equipment shown here uses many sensors attached to the patient. The signals from these must be amplified before they can be used.

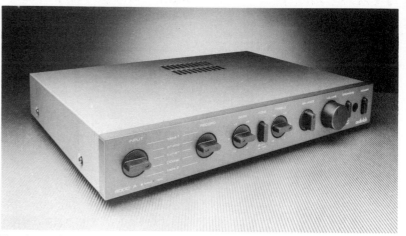

The signals from tape recorders, disc players and radio tuners all need amplifying before they can drive a speaker.

Radio transmitters and receivers both need amplifiers to increase signals, especially when communicating with satellites. This is British Telecom's satellite station at Madley.

Summary

- When a small input signal goes into an amplifier a larger signal is produced at the output.
- The ratio of the output signal compared to the input signal is called the **gain** of the amplifier.
- Amplifiers with negative gain produce an output signal which is out of phase with the input signal.
- When two amplifiers are cascaded together the overall gain can be found by multiplying their two gains together.
- Clipping occurs when the output signal tries to exceed the power supply.

Questions

1 Explain in your own words the terms printed in darker type: 'Two **amplifiers** can be **cascaded** together. Care must be taken to prevent **clipping** by keeping the **gain** down to a sensible size.'.
2 If an input signal of 25 mV is applied to an amplifier with a gain of 60, what is the amplitude of the output signal? (Assume that there is no clipping.)
3 An amplifier is run from a 12 V power supply. What is the largest input signal that can be applied before clipping occurs?
4 Two amplifiers each have a gain of 20. What is the overall gain when they are cascaded together?

41
Transistor Amplifiers

In chapter 8 the transistor was used as an electronic switch. In this chapter we shall see how it may be used to amplify signals.

In the transistor switch a small change in the voltage at the base (from 0 V to 0.7 V) turned the transistor on and produced a large change in the voltage at the collector (9 V down to 0 V). (Figure 41.1)

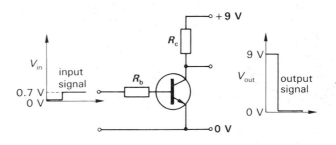

Figure 41.1 Voltage changes around a transistor used as a switch

To use the transistor as an amplifier we send a small current into the base to turn the transistor 'halfway on', making the collector sit at 4.5 V. This is called **biassing** the transistor (Figure 41.2). Then, any further small change to the voltage at the base will produce a much larger change in the voltage at the collector as the transistor is driven further on or off.

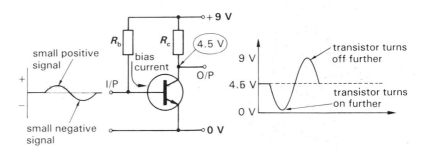

Figure 41.2 Voltage changes around a biassed transistor

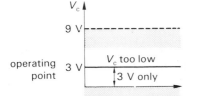

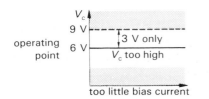

Figure 41.3 Incorrect biassing reduces the possible size of the output signal

Importance of correct biassing

If the transistor is correctly biassed the collector voltage should be able to swing an equal amount either side of its **operating point**: up by 4.5 V to +9 V or down by 4.5 V to 0 V.

The result of **incorrect biassing** is shown in Figure 41.3.

Improved biassing

The biassing arrangement shown in Figure 41.2 is not usually used in practical circuits because it is unstable and too dependent on choosing exactly the right component values to set the operating point correctly. A better arrangement is shown in Figure 41.4.

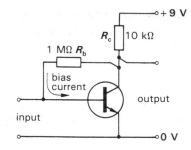

Figure 41.4 Transistor with stable biassing arrangement

Connecting to a signal source

If a signal source (say a moving-coil microphone) was connected directly to the circuit shown in Figure 41.4, the bias current would flow through the coil of the microphone and not into the transistor. This would upset the biassing.

Coupling capacitor

The answer is to connect a capacitor (which lets signals pass but does not conduct direct currents) between the signal source and input. (See Figure 41.5.) A capacitor is also connected between the output and the load. A capacitor used in this way is called a **coupling capacitor** or a **blocking capacitor** because it blocks the bias current.

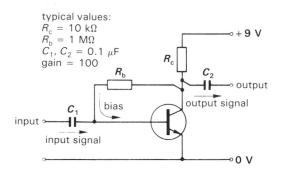

Figure 41.5 Practical amplifier circuit

Summary

● A transistor can be used as a voltage amplifier.
● It must be correctly biassed by sending a small current into the base.
● When the transistor is biassed at the right operating point the collector voltage is half the power supply voltage.
● Coupling capacitors must be used at the input and output.

Questions

1 Explain in your own words the terms printed in darker type: 'Before a transistor can be used as an amplifier it must be correctly **biassed**. At the correct operating point the **collector voltage** is half the **supply voltage**. When the amplifier is connected to a **signal source**, then **coupling capacitors** must be used.'.
2 If a transistor amplifier has a gain of 100 and works off a 12 V supply, calculate:
   (a) the collector voltage if the transistor is correctly biassed
   (b) the maximum change in collector voltage that a signal can produce
   (c) the maximum input signal that can be applied before clipping occurs.

42
More About Transistors

In the last chapter we saw two ways of biassing a transistor. One of these was more stable. We shall now see why. First of all we must go a little deeper into the way transistors behave.

Gain of a transistor

The transistor we have looked at should be called a **bipolar transistor**. The current flowing through the collector (I_c) is controlled by the current flowing into the base (I_b).

The **current gain** of the transistor can be found by:

$$\text{current gain} = \frac{I_c}{I_b}$$

This is about 200 for most small signal transistors like a BC108.

The problem is that the current gain of a transistor can change. (A rise in temperature could do this.) What effect will a change in current gain have on the two circuits shown in Figures 42.1 and 42.2? Remember that the idea of biassing is to cause just enough current to flow through R_c to make the collector voltage 4.5 V.

Unstable biassing

In the circuit shown in Figure 42.1 the base current comes straight from the +9 V rail through R_b. If the current gain increases then the collector current will rise and the voltage at point C will drop. Not only will this upset the operating point but the increased current will heat up the transistor even more, making things worse! This is known as **thermal runaway**.

Stable biassing

The base current comes from point C (Figure 42.2). If the current gain should increase, the voltage at C will drop as before. But now, this will mean that less base current will flow and so the collector current will be reduced. This will mean that the voltage at point C will rise again as the transistor turns off further. This is clearly a much more stable way of biassing. This is an example of **negative feedback** (see chapter 45).

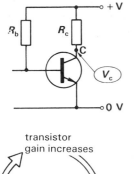

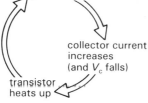

Figure 42.1 Unstable biassing can produce thermal runaway

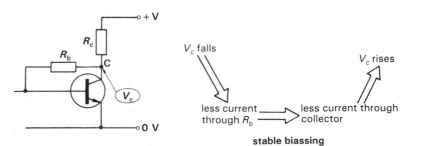

Figure 42.2 Stable biassing

An even more stable transistor amplifier

In the circuit in Figure 42.3, the bias current is taken from the fixed voltage in the middle of a potential divider. This is even more stable.

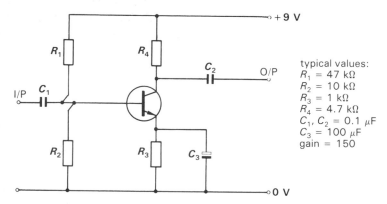

Figure 42.3 Transistor amplifier with fully stabilised operating point

Resistors R_1 and R_2 fix the base voltage. Since this will be about 1.5 V we need another resistor (R_3) to let the emitter voltage rise slightly and keep the base voltage at 0.7 V above the emitter. The capacitor C_3 lets a.c. signals bypass R_3 so that, as far as they are concerned, the emitter is still at 0 V. The gain of the whole circuit will be about 150.

Input resistance

In all of the amplifier circuits we have seen so far there is one major problem; the **input resistance** is fairly small. This means that a large current will flow into the input and this may reduce the p.d. of the input signal, especially if it comes from a device like a crystal microphone. (See chapters 31 and 35.)

Field-effect transistors

The **field-effect transistor** (or **FET**) (Figure 42.4) behaves in a different way to the bipolar transistors we have seen so far. It has a very high input resistance (at least 10 MΩ) and so it takes almost no current at all from a signal source. In effect it is a true **voltage amplifying** device.

How an FET behaves

If the gate is given a **negative** voltage, then there is a large resistance between the drain and the source and no current can flow down through R_D into the FET.

If we steadily reduce the negative p.d. on the gate then the resistance between the drain and the source decreases. This allows a current to flow through R_D and so the voltage at D drops.

Note that almost no current can flow into the gate.

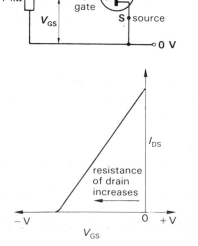

Figure 42.4 Circuit to investigate the behaviour of a field-effect transistor and typical results obtained

110

FET amplifier

In the circuit shown in Figure 42.5 the FET is being biassed by the resistor R_S. This produces a p.d. between G and S and lets some current flow through R_D and the FET. The input signal goes through C_1 and the change in voltage produces a change in current through R_D. This produces an amplified voltage signal at D. The gain is about 20. Notice that a $+18\,V$ power supply is needed.

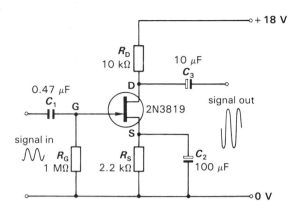

Figure 42.5   Voltage amplifier using an FET

Summary
- The current gain of a bipolar transistor is equal to $I_b//I_c$.
- Current gain can change, especially if the temperature rises.
- In an unstable amplifier this will lead to thermal runaway.
- In a stable amplifier the effects of a rise in gain will be evened-out by the reduced base current.
- An even more stable amplifier fixes the base current by fixing the voltage at the base using a potential divider.
- A field-effect transistor (FET) has a high input resistance and takes very little current from the signal source.
- An FET amplifies changes in voltage. A bipolar transistor really amplifies changes in current.

Questions
1 How can we calculate the current gain of a transistor?
2 If a transistor has a current gain of 80, how much collector current will flow if 0.5 mA flows into the base?
3 What can make the gain of a transistor rise?
4 What is meant by thermal runaway?
5 Explain what happens to the collector voltage when the current gain rises in a stable amplifier.
6 What will happen to the collector voltage if the current gain falls?
7 Draw a table to show the difference between a bipolar and field-effect transistor in terms of:
   (a) input resistance
   (b) power supply requirements
   (c) uses.

111

43 Operational Amplifiers

Operational amplifiers (or **op-amps**) are one of the most useful building blocks. They can be bought as ICs. The **741** (used as a comparator in chapter 10) is an op-amp. They have:

- A very high **gain** (about 200 000).
- A very high input impedance (about 10 MΩ). This means that very little current flows from a signal source into the op-amp.
- A low output impedance (about 2 kΩ for a 741). This means that it can be easily connected to other stages in a system.
- Two inputs. The op-amp amplifies the **difference** between the voltages put on these two inputs.

How the op-amp behaves

The op-amp amplifies the **difference** between the voltages which we put on the two inputs. (See Figure 43.1.) This is simplified if we fix one of the inputs at 0 V (by connecting it to the 0 V rail) and put a signal onto the other input. This can be done using the non-inverting input or the inverting input.

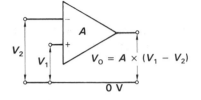

Figure 43.1   General behaviour of an operational amplifier (op-amp)

Using the non-inverting input *(Figure 43.2)*

output signal $= A \times$ **input signal**

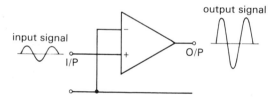

Figure 43.2   Using the non-inverting input. (Note that the input and output signals are not drawn to scale.)

Using the inverting input *(Figure 43.3)*

output signal $= -A \times$ **input signal**

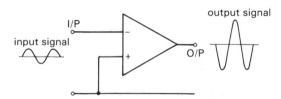

Figure 43.3   Using the inverting input. (Note that the input and output signals are not drawn to scale.)

(Note the minus sign; this simply means that the output voltage is negative if the input signal is positive and vice versa.)

Using op-amps

The op-amp is never used in the simple way shown here. Because the gain of an op-amp is so large, any reasonable input signal would be amplified so much that **clipping** would occur.

Another problem is that the gain changes and is affected by changes in supply voltage, temperature and frequency of the signal being amplified. This means that we have to connect up op-amps in such a way that any change in the gain does not cause problems.

Op-amp circuits

There are two main ways of using op-amps as signal amplifiers: the inverting amplifier and the non-inverting amplifier.

Inverting amplifier *(Figure 43.4)*

This time the gain is a more sensible size (it actually depends on the size of the two resistors R_f and R_i and is equal to $-R_f/R_i$). A typical calculation is shown opposite. Because the values of resistors do not change very much the gain of this circuit will stay steady (or **stable**).

if R_f = 100 kΩ
and R_i = 10 kΩ

then:

gain $= -\dfrac{100}{10}$

$= -10$

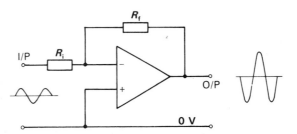

Figure 43.4 Op-amp used as an inverting amplifier

Non-inverting amplifier *(Figure 43.5)*

Once again the gain depends on the size of R_f and R_i (this time it is equal to $1 + R_f/R_i$). This is also stable.

if R_f = 100 kΩ
and R_i = 10 kΩ

then:

gain $= 1 + \dfrac{100}{10}$

$= 1 + 10$

$= 11$

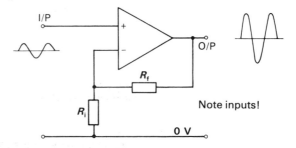

Figure 43.5 Op-amp used as a non-inverting amplifier

113

Useful terms

The resistor R_f is often called the **feedback resistor**. The resistor R_i is often called the **input resistor**.

Advantages of using op-amps

● The high input impedance of the op-amp means that in effect the input resistance of the inverting amplifier (Figure 43.4) is equal to R_i. This lets you choose the best value for R_i to match the amplifier to the signal source.
● The cost and size of an amplifier built around an op-amp are less than one using transistors. It is also more reliable.
● It is quite easy to design systems using op-amps. The next four sections show some examples of this:

Voltage follower *(Figure 43.6)*

The **voltage follower** is simply an amplifier with a gain of 1. This means that the output signal is exactly the same as (or 'follows') the input signal. Although this may seem pointless the circuit is useful. Because op-amps have a very high input impedance and a low output impedance the voltage follower is a way of matching a high impedance signal source with a low impedance load.

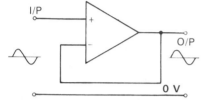

Figure 43.6   A voltage follower circuit

Inverter *(Figure 43.7)*

The **unity gain inverter** is an inverting amplifier with a gain of -1. Like the voltage follower it does not increase the amplitude of a signal. The important point here is that, because the op-amp has such a very high gain, the signal arriving at point X need only be very small. In fact the voltage at point X will be almost 0 V. (Point X is called a **virtual earth**.)

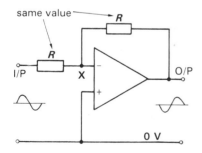

Figure 43.7   The unity gain inverter circuit

Signal mixer

A **signal mixer** produces an output which is the sum of two or more input signals. The circuit shown in Figure 43.8 is an example of a signal mixer which can be used to mix two audio signals. VR_1 and VR_2 act as volume controls for each signal. Because the '−' input in an inverting amplifier is a virtual earth, when these signals arrive at point X they are so small that there is no chance of, say, input signal 1 going back through R_{i2} and being affected by VR_2.

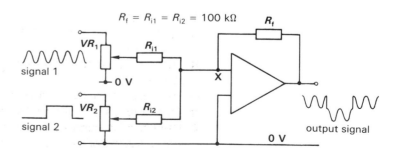

Figure 43.8   Circuit for a simple two-input signal mixer

114

Ramp generator (or integrator)

The ramp generator turns a **step voltage** signal into a **ramp** (Figure 43.9). It will also turn a square wave signal into a sawtooth signal (see chapter 29). It is used in the timebase of an oscilloscope (see chapter 48), in televisions and in A to D converters (chapter 27).

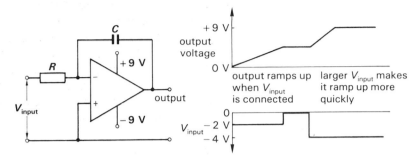

Figure 43.9 A ramp generator or integrator circuit

Power supplies

Most op-amps (like the 741) need two power supplies (Figure 43.10). One supply must be connected with '+' to pin 7 ('−' to 0 V rail) and the other is connected with '−' to pin 4 ('+' to 0 V rail).

Provided both supplies are the same size, any voltage between 3 V and 18 V will work (9 V batteries are a good method).

Note that there is no direct connection from the op-amp to the 0 V rail.

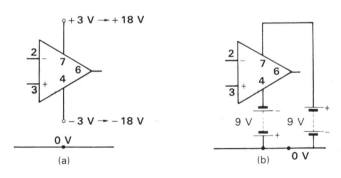

Figure 43.10 Power supply arrangements for an op-amp   (a) General arrangement   (b) Using two 9 V batteries

Clipping

As in any amplifier, if the output signal is too large then clipping may happen. With two 9 V batteries as a power supply, clipping will happen when the output signal reaches about 8 V.

Summary

- Op-amps have a very large gain, a high input impedance and a low output impedance.
- They amplify the difference between voltages at their two inputs.
- They are normally used with two resistors in the circuit.
- This reduces the gain and makes it more stable.
- They can be used as amplifiers, for impedance matching, as a signal mixer, or as a ramp generator.

Questions

1 What is 'op-amp' an abbreviation of?
2 Look at the circuit in Figure 43.11.

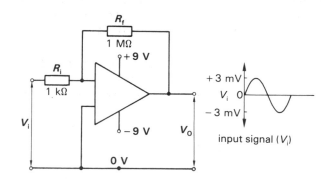

Figure 43.11

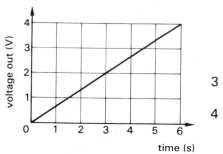

Figure 43.12

(a) How big is the gain of this amplifier?
(b) How big is the output signal?
(c) If the **input signal** was increased to 6 mV, how big would the output signal be?
(d) What would happen if the power supply to the op-amp was only 5 V?

3 Why do we use op-amps in circuits like the one in Figure 43.11 and not just on their own (without R_f and R_i)?

4 The output from a ramp generator like the one in Figure 43.9 is shown on the graph in Figure 43.12. The ramp rate (in volts/ second) describes how quickly the output voltage rises. It can be found by the formula:

Ramp rate $= -V/RC$ where V is the input voltage.

(a) Calculate the ramp rate shown on the graph.
(b) Calculate the value of the capacitor (C) if the input voltage was -5 V and $R = 1$ MΩ.
(c) Draw a graph of the output voltage if a -10 V input was applied for 3 seconds.

44
Power Amplifiers

Power amplifiers produce a large voltage signal at their output which drives a large current through the load (e.g. a speaker). A high gain voltage amplifier operating on a high voltage power supply produces a large voltage signal. This is followed by a current amplifier which supplies the large current required. See Figure 44.1.

Power transistors

A large current will produce **heat**. Current amplifiers use special transistors which can stand the heat produced without getting too hot and failing. These are called **power transistors** or sometimes **output transistors**.

Heat sinks

To make sure that power transistors do not overheat they are generally mounted on a large piece of aluminium called a **heat sink**. This **dissipates** the heat by taking the heat away from the transistors into the air. The fins help the dissipation.

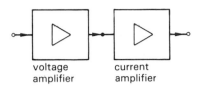

voltage amplifier   current amplifier

Figure 44.1   General arrangement of a power amplifier

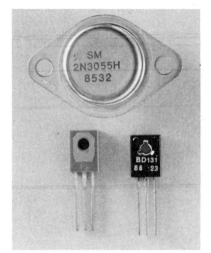

Typical examples of power transistors.

Typical examples of heat sinks.

Matching the load to the amplifier

The power amplifier has a low **output impedance**. This means that it must be connected to a load with a similarly low impedance to make sure that the maximum amount of power is produced in the load. This is called **matching** the load to the amplifier. Figure 44.2 shows why matching is important.

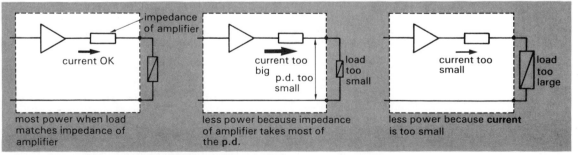

impedance of amplifier

current OK

most power when load matches impedance of amplifier

current too big
p.d. too small
load too small

less power because impedance of amplifier takes most of the p.d.

current too small
load too large

less power because **current** is too small

Figure 44.2   The amplifier and load must be correctly matched

117

The amplifier in a car radio draws a large current from the car battery. Could this radio be used with normal batteries?

Class A amplifier circuit

The simplest circuit for a **class A amplifier** is shown in Figure 44.3. Two simple amplifiers are cascaded together. A power transistor is used in the second stage.

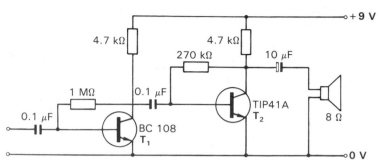

Figure 44.3   A two-stage amplifier with a 'class A' output stage

The output transistor (T_2) is biassed so that current flows through it all the time, even if no signal is going into the input. This is a feature of all class A amplifiers. If batteries are used to power the circuit they would soon run flat. Class A amplifiers are usually only used with mains-operated power supplies or in car radios (where the battery is recharged).

Class B amplifier circuit

The class B amplifier is also called a **push-pull** amplifier (Figure 44.4). Two output transistors are used. Each one is slightly biassed on.

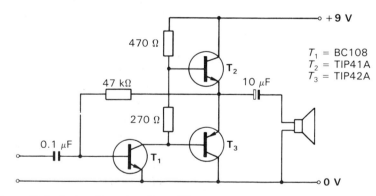

Figure 44.4   A class B, push-pull amplifier

Current flows only when an input signal is applied. Transistor T_2 only conducts when the input signal is a positive voltage. Transistor T_3 conducts only when the input signal is a negative voltage. This arrangement, with one transistor **pushing** and the other **pulling**, is much more efficient.

Notice that T_3 is a p-n-p type transistor. (An n-p-n type would 'blow' if a negative voltage was put on its base.)

Integrated circuit power amplifiers

Not surprisingly there are several integrated circuits which are designed to be used as power amplifiers. They range from a few watts up to 100 W or more.

Typical circuit

Manufacturers always supply detailed **data sheets** on how to use their particular IC. A typical circuit using the LM380 is shown in Figure 44.5. Notice that very few extra components are needed.

A selection of integrated circuit power amplifiers. Note the fins which help to get rid of the heat quickly.

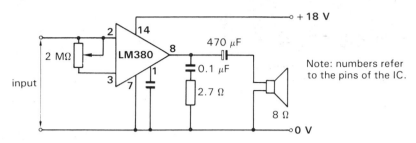

Figure 44.5   A 2 watt amplifier using an LM380

Summary

● A power amplifier produces a large voltage and current.
● Power transistors which can stand a large current are used at the output stage.
● A heat sink is used to dissipate the heat produced.
● The amplifier and load must be matched to produce maximum power output in the load.
● Class A amplifiers use current all the time.
● Class B amplifiers only draw current when a signal is present.

Questions

1   Explain in your own words the terms printed in darker type: 'A **power amplifier** produces a large voltage signal. When connected to a **load** which **matches** the amplifier a large current flows into the load. **Power transistors** are used with a **heat sink** at the **output stage**.'.
2   Draw the part of the circuit in Figure 44.5 which amplifies the voltage of the input signal.
3   Why do batteries last longer when used with class B amplifiers rather than with class A amplifiers?
4   Speaker leads must be kept in good condition. What might happen if the two wires accidentally touched together?
5   Look through some catalogues and find an integrated circuit power amplifier that can deliver about 8 W. Draw a circuit to show how it might be used.

45
Feedback

The idea of **feedback** is to take a sample of the signal going out of a system and feed this sample back into the input. The reason for this is to allow the system to adjust itself. Systems which use feedback are called **closed-loop systems**. Systems without any feedback are called **open-loop systems**.

Feedback in amplifiers

Feedback is a very important part of many electronic systems. It is particularly useful in amplifiers. To explain this, first consider an amplifier without any feedback (i.e. **open-loop**).

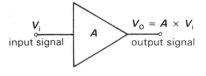

Figure 45.1 An amplifier without feedback is an open loop system

Amplifier without feedback *(Figure 45.1)*

The **output signal** of the amplifier is simply $A \times$ **input signal**. However, the actual gain can easily change with changes in temperature, age of components, frequency of signal, power supply variation, etc. This does not make a very stable system.

Amplifier with feedback *(Figure 45.2)*

Suppose we now feed some (not all!) of the output signal back into the input. (Simply using a resistor is often good enough.) What actually goes into the amplifier is a combination of the input signal and the feedback signal (i.e. a **closed-loop system**).

What happens now depends on how the input and feedback signals combine.

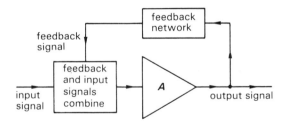

Figure 45.2 An amplifier with feedback is a closed loop system

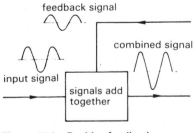

Figure 45.3 Positive feedback

Positive feedback *(Figure 45.3)*

If the feedback signal and the input signal add together then the signal going into the amplifier is larger. We call this **positive feedback**. It increases the gain but makes it unstable. Positive feedback is not used in amplifiers but it is very important in oscillators where it will keep oscillations going.

Negative feedback *(Figure 45.4)*

If the feedback signal subtracts from the input signal then the signal going into the amplifier is smaller. We call this **negative feedback**. It reduces the gain but makes it more stable.

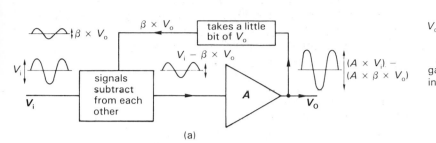

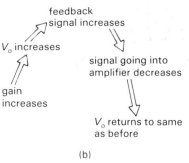

(a)

(b)

Figure 45.4 Negative feedback
(a) Block diagram showing the
reduction in gain (b) Why the system
is stable

Explanation

The signal into the input is $(V_i - \beta V_0)$. This is amplified so that the signal from the output is $(A \times V_i - A \times \beta V_0)$. If the gain (A) should increase then both the **amplified input signal** $(A \times V_i)$ and the **amplified feedback signal** $(A \times \beta V_0)$ increase. Since these two subtract from each other the two effects cancel out. The output signal stays stable.

Negative feedback in op-amp circuits

Figure 45.5 shows the inverting amplifier circuit. This uses negative feedback to reduce the high gain of the op-amp (from about 200 000) to a lower but more stable value $(R_f/R_i = -10)$. The input signal goes into the '−' input. If this is a positive voltage the output signal will be a negative voltage. The resistor R_f feeds some of this negative voltage back into the '−' input. The result is to reduce the signal actually going into the op-amp.

Figure 45.6 shows the non-inverting amplifier circuit. This time the output voltage is positive if the signal is positive. The feedback signal is sent into the '−' input, however. Since the op-amp amplifies the difference between these two signals we still have negative feedback.

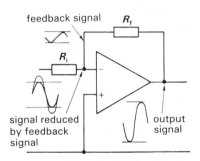

Figure 45.5 Negative feedback in the op-amp used as an inverting amplifier

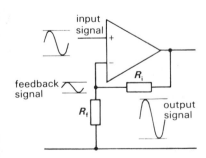

Figure 45.6 Negative feedback in the op-amp used as a non-inverting amplifier

Summary

● Feedback consists of feeding a sample of the output signal back into the input of a system.
● If the feedback signal subtracts from the input signal we have negative feedback. This reduces the gain but makes it more stable.
● If the feedback signal adds to the input signal we have positive feedback. This increases the gain but makes it unstable.

Questions

1 What is meant by a closed-loop system? Give an example.
2 What are the two effects of negative feedback on gain?
3 What are the two effects of positive feedback on gain?
4 How is negative feedback produced in Figure 45.6?

46
Filters

Filters are **frequency-dependent** circuits. This means the behaviour of the circuit depends on the frequency of the signal going in. The behaviour of a filter is easiest to describe by drawing a **frequency response** diagram which shows how the output changes as the frequency of an input signal increases. For example:

High pass filter (*Figure 46.1*)
A high pass filter lets high frequency signals pass through easier than low frequency signals.

Figure 46.1 High pass filter: symbol and frequency response curve

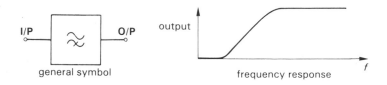

Low pass filter (*Figure 46.2*)
A low pass filter lets low frequency signals pass through easier than high frequency signals.

Figure 46.2 Low pass filter: symbol and frequency response curve

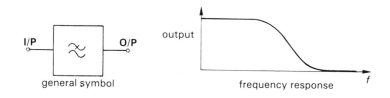

Band pass filter (*Figure 46.3*)
A band pass filter lets signals inside a certain range pass through easier than all other signals.

Figure 46.3 Band pass filter: symbol and frequency response curve

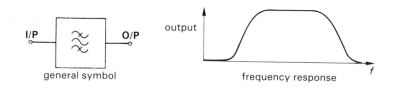

Components used in filters
Two components are useful in filters: **capacitors** pass high frequency signals while **inductors** pass low frequency signals.

Some examples of filters

● A **band pass filter** is used in the **tuning stage** of a radio, where all stations are filtered out except the one being listened to. (Figure 46.4)

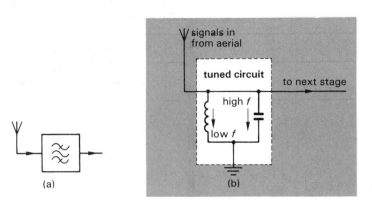

Figure 46.4 Tuning stage in a simple radio   (a) General arrangement (b) Typical circuit. Usually a variable capacitor is used.

● **High and low pass filters** are used in a loudspeaker crossover. The bass and treble signals are each sent to their own speaker. (Figure 46.5)

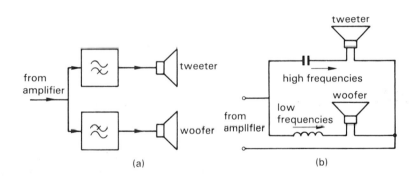

Figure 46.5 Loudspeaker crossover system   (a) General arrangement (b) Typical circuit

● A **suppressor** is a **high pass filter** connected between the live and earth wires in electrical appliances. It takes away the sudden electrical pulses produced when, say, a washing-machine turns on or off, and prevents interference to other circuits. (Figure 46.6)

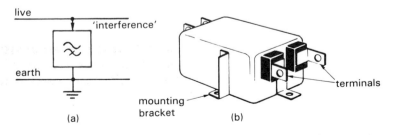

Figure 46.6 Suppressing mains-borne interference   (a) General arrangement (b) Example of a typical mains suppressor device

123

● **Tone controls** in audio equipment use filters to separate the bass and treble. A **graphic equaliser** contains many **band pass filters**, each one tuned to a particular frequency range.

A graphic equaliser is used to cut or boost particular frequencies. This can help to correct defects in the acoustic properties of the room or speaker cabinets.

Simple tone control

The circuit shown in Figure 46.7, connected before a power amplifier, will allow treble or bass to be cut from the final signal.

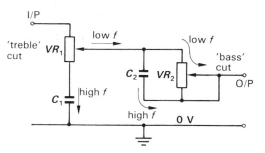

Figure 46.7   Simple passive tone control

How it behaves

1   When the sliders of VR_1 and VR_2 are at the top, all signals pass straight through unaffected.
2   If the slider of VR_1 is moved down, it becomes easier for high frequencies to pass to earth through C_1 rather than carry on through the rest of the circuit. Treble is cut from the signal.
3   If the slider of VR_2 is moved down, low frequency signals have to travel through more resistance and are reduced. Bass is cut from the signal. High frequencies are unaffected because they can go through C_2.

Active filters

An **active filter** can amplify certain frequency signals as well as reduce them. Active filters can be made with op-amps. An example of an active bass control is shown in Figure 46.8. Moving *VR* down will produce more feedback resistance. This will increase the gain of the amplifier (gain = R_f/R_i). High frequency signals bypass *VR* through C_1 and C_2 and are unaffected.

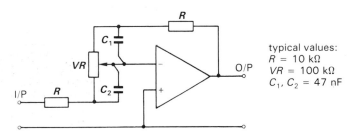

typical values:
R = 10 kΩ
VR = 100 kΩ
C_1, C_2 = 47 nF

Figure 46.8 An active bass control circuit

Summary

- Filters are frequency-dependent circuits.
- They are termed high pass, low pass or band pass depending on which frequencies they allow through.
- Filters are used in tuning a radio, speaker systems, tone controls and to suppress interference.
- Passive filters reduce certain frequencies while active filters can also amplify certain frequencies.

Questions

1 Draw frequency response graphs for high, low and band pass filters. Give an example of how each one might be used.
2 Draw a circuit for a speaker crossover and explain how it works.
3 Explain how a simple tuned circuit like the one shown in Figure 46.4 will allow only signals of a certain frequency to pass out of it. What happens to the other signals?

47
Signal Generators

A **signal generator** is a piece of equipment used in testing circuits and systems. It can produce signals (usually over a range of frequencies) whenever we want them. It contains at least one **oscillator**.

An oscillator is a circuit which can produce signals on its own without any kind of input. We have already seen one kind of oscillator, the astable.

Types of signal generator

Signal generators are generally described by the **frequency** of the signal they produce and its **waveform** (e.g. **sine wave**, etc.).

An **audio frequency** (or a.f.) oscillator produces signals which could be turned into sounds that we can hear by a speaker (i.e. somewhere between about 20 Hz and 20 kHz).

A **radio frequency** (or r.f.) oscillator produces signals of the same range of frequencies as radio waves (see chapter 57) (i.e. somewhere between about 3 kHz and 300 GHz).

A typical signal generator used in schools. What are the highest and lowest frequency signals that it can produce?

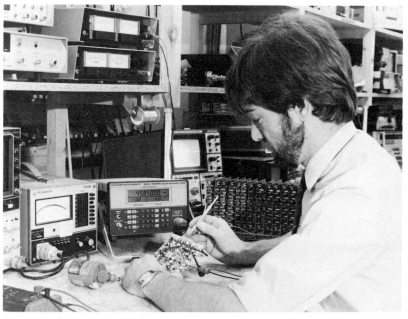

A well-equipped test bench uses at least one signal generator. The one shown here (in the middle) can produce signals up to 1 GHz. What other test equipment do you recognise?

48
Oscilloscopes

An oscilloscope (or **scope**) is a useful item of test equipment. It is most often used to 'look' at electrical signals.

A typical portable oscilloscope. This one has two Y-inputs so that two signals can be compared. Compare this scope with the one that you use. Try to identify all the controls.

Testing part of the telephone system. The engineer can see exactly what signals are being produced. The large oscilloscope screen allows a lot of detail to be shown.

The screen of the scope works in much the same way as a TV screen (see chapter 60), using an electron beam to make the screen phosphor glow. Controls are provided to adjust the **intensity** and **focus** of the spot of light produced. The spot can be moved along the **x-axis** (i.e. horizontally) and along the **y-axis** (i.e. vertically).

Two kinds of measurements are possible: **voltage** and **time**.

Voltage measurement

The voltage to be measured is connected to the **y-input**. A positive voltage makes the spot move upwards; negative voltages make the spot move down. To find the value of the voltage we measure the amount of movement using the grid drawn on the screen and multiply this by the setting on the **sensitivity** control.

An example of measuring a voltage is shown in Figure 48.1:

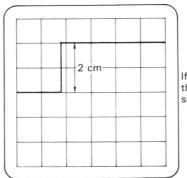

2 cm

If sensitivity = 5 V/cm
then this step voltage
signal is   5 × 2
             = 10 V

Figure 48.1   Measuring a change in voltage

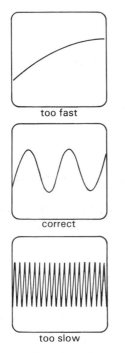

too fast

correct

too slow

Figure 48.2 The effect of the timebase speed on the trace

Looking at signals

Although the scope could be used as a simple voltmeter it is more useful for measuring changing voltages (signals). A circuit called the **timebase** moves the spot sideways at a steady rate and flies back when it reaches the right-hand side to start again from the left. The speed can be adjusted by the **timebase control**.

If a signal is connected to the y-input, the screen will show a **trace** of how the voltage of the signal varies with time. We may need to adjust the timebase to get a clear trace (see Figure 48.2).

Measurement of time and frequency

Since the timebase moves the spot at a known speed, the x-axis can be used as a clock. An example will illustrate this (Figure 48.3):

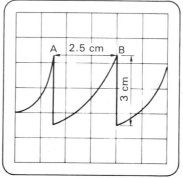

sensitivity = 5 V/cm
timebase = 2 ms/cm

Figure 48.3 Example of a trace. The sensitivity and timebase speed controls were set as shown.

Time

The **time** taken for the signal to go through one complete cycle (from A to B) is found by:

$$\text{time} = (\text{distance A–B}) \times (\text{timebase setting})$$

In this example (*Figure 48.3*):

$$\text{time} = 2.5\,\text{cm} \times 2\,\text{ms/cm}$$
$$= 5\,\text{ms}$$

Frequency

The **frequency** of the signal can be found by calculating how many cycles can occur in one second. In other words:

$$\text{frequency} = 1/\text{time for one cycle (s)}$$

In this example (*Figure 48.3*):

$$\text{frequency} = 1/5\,\text{ms}$$
$$= 1/0.005\,\text{s}$$
$$= 200\,\text{Hz}$$

Amplitude

The **amplitude** of the signal is found by:

amplitude height (h) × sensitivity

In this example *(Figure 48.3)*:

$$\text{amplitude} = 3\,\text{cm} \times 5\,\text{V/cm}$$
$$= 15\,\text{V}$$

Uses of oscilloscopes

Apart from the simple measurements we have looked at, the oscilloscope can be very useful for:

● Checking the output of an amplifier against the signal going in. The scope will show if there is any **distortion** in the signal coming out.
● Testing signal sources. The input on an oscilloscope has a very high impedance and so it takes very little current from the signal source.
● Testing the frequency response of audio systems. A signal generator can provide a range of input signals. The oscilloscope can give a much better indication of the amplitude of the output than our ears can provide.

Summary

● An oscilloscope can be used to measure voltage or time.
● To measure voltage we look at the vertical scale on the screen.
● To measure time we look at the horizontal scale on the screen.
● We can find the frequency of a signal by measuring the time for one cycle and then calculating how many cycles can fit into one second.

Checking an amplifier with a dual beam scope. The input and output signals can be easily compared.

49

Exercises

1 (a) What is the purpose of an amplifier?
(b) Name three devices or circuits that could be connected to the input of an amplifier.
(c) Name three devices or circuits that could be connected to the output of an amplifier.

2 The amplifier shown in Figure 49.1 has a gain of 20 and is powered from a +9V supply. Input and output signals are as shown:

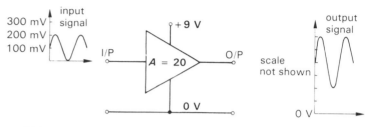

Figure 49.1

(a) (i) What is the amplitude of the input signal shown?
(ii) What is the amplitude of the output signal shown?
(iii) What would the amplitude of the output signal be if the input signal were increased to 300 mV?
(b) A second stage with a gain of −2 is now added (Figure 49.2):

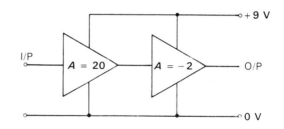

Figure 49.2

(i) What does a negative gain indicate? (ii) What is the overall gain of the system? (iii) Draw a diagram showing the output signal from this system.

3 Copy the circuit in Figure 49.3 for a transistor amplifier and include the following: (i) coupling capacitors (ii) a microphone as an input (iii) a speaker as an output.

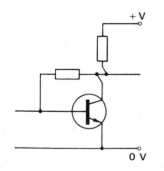

Figure 49.3

4   Figure 49.4 shows three ways to bias a transistor:

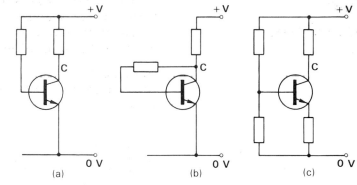

Figure 49.4

(a) Which method produces the most stable arrangement?
(b) When correctly biassed, what should the voltage at point C be?
(c) What will happen to the voltage at point C in circuit (a) if the gain of the transistor increases for some reason?
(d) What effect will this have in circuits (b) and (c)?
(e) Which of these circuits use negative feedback?

5   The signal shown in Figure 49.5 is sent into three different amplifiers (A, B and C) in turn. The output signal for each one is shown.

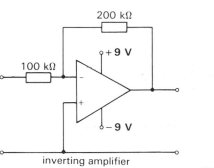

inverting amplifier

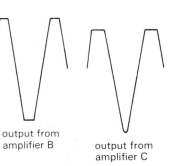

Figure 49.5

(a) Which amplifier is probably working correctly?
(b) (i) What is the name of the distortion shown from amplifier B?   (ii) How is this distortion caused?   (iii) How can it be reduced?
(c) What is the reason for the distortion shown in: (i) amplifier A   (ii) amplifier B?

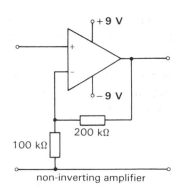

non-inverting amplifier

Figure 49.6

6   Two ways of using an op-amp are shown in Figure 49.6.
(a) What is the gain of each circuit?
(b) What would the output voltage be in each case if: (i) the input voltage was 2 V   (ii) the input voltage was 4 V?
(c) How could we most easily vary the gain of each circuit?

131

Section 6

Power Supplies

50
Introduction to Power Supplies

The purpose of a **power supply** is to provide the e.m.f. which will drive the current around the system and to provide the energy that will finally be turned into something else at the output.

In many systems the power supply is the largest, heaviest and perhaps most expensive part of the system.

Ways of supplying power

There are several ways of supplying power to a system:

- **Cells** or **batteries** are useful for portable equipment. They cannot supply much current and can be expensive.
- The **240 V a.c. mains** can supply more current than batteries. A **mains unit** is needed to turn the a.c. mains into the low-voltage d.c. supply needed by most systems.
- **Solar cells** can give a constant supply of power. They cannot give much current (unless they are very large) and work only when the sun is shining!
- **Fuel cells** work by a chemical reaction, usually at a high temperature. They are mainly used in spacecraft.

Solar cells can be used for more than just calculators. A large cell like this one can recharge a car battery.

Hardly pocket-sized! These are the fuel cells of an Apollo 2TV-1 spacecraft.

132

The ideal power supply

A power supply provides a steady e.m.f. When connected to a system, current flows out of the power supply. The amount of current depends on the demands of the system. A perfect power supply would be able to provide an e.m.f. which never changes no matter how much current flows from it. It would be able to provide unlimited amounts of energy both to run the circuits in the system and be converted into other forms of energy at the output. Nobody has ever designed the perfect power supply!

Real power supplies

Power supplies are made from conductors, components and so on. These have **resistance**. Any current flowing has to flow through this **internal resistance** before it gets out into the rest of the system. This causes two problems if we try to take too much current:

1 Heat is produced inside the power supply. This can damage the power supply.
2 Some of the e.m.f. is used inside the power supply so that less is available for the rest of the system. See Figure 50.1.

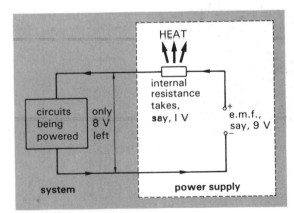

Figure 50.1   A real power supply loses some of the e.m.f. in its internal resistance

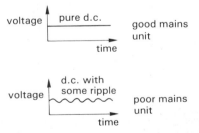

Figure 50.2   Ripple in the e.m.f. from a mains unit

Problems with mains units

A perfect mains unit produces a completely **smooth** d.c. from the a.c. mains. Poor mains units have a small **ripple** in their e.m.f. left over from the a.c. going in. (See Figure 50.2.) This can be reduced by good design.

133

Summary

● The power supply provides the e.m.f. for the system. It must supply enough energy to run the system and the final output.
● Batteries, the mains, solar cells and fuel cells can be used as power supplies. All have advantages and disadvantages.
● Internal resistance limits the amount of current (and energy) that a power supply can give.
● Internal resistance also reduces the e.m.f. available for the system when a large current is being taken.

Questions

1   What are the four common types of power supplies?
2   Describe one way in which batteries are better than a power supply which uses the mains.
3   Describe one way in which a power supply which uses the mains is better than batteries.
4   Look at the circuit in Figure 50.3.
   (a) Calculate the total resistance in the circuit.
   (b) Calculate how much current flows.
   (c) Calculate the p.d. lost across the internal resistance.
5   The p.d. produced by a simple (unstabilised) power supply was measured while the current drawn from it was increased. The following results were obtained:

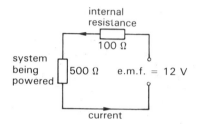

Figure 50.3

| output p.d. (V) | 12.0 | 11.5 | 11.1 | 10.6 | 10.1 | 9.5 | 9.1 |
| current drawn (A) | 0 | 0.4 | 0.8 | 1.2 | 1.6 | 2.0 | 2.4 |

(a) What causes the output voltage to fall like this?
(b) Draw a graph of output voltage against current drawn.
(c) From your graph, how much current can be drawn before the voltage drops to below 10.5 V?
(d) What is the output voltage of the supply when 1.0 A is drawn?
(e) Calculate how much p.d. is lost inside the power supply when 1.0 A is drawn.
(f) Hence, calculate the power being dissipated inside the supply when 1.0 A is drawn.
(g) If the power supply can safely dissipate up to 5 W, estimate the maximum safe current that can be drawn from the supply.

51
Batteries and Cells

A **cell** is a single source of power. It is made from two different conductors (or **electrodes**) separated by a chemical (or **electrolyte**). The general design of a typical cell is shown in Figure 51.1. Most cells provide an e.m.f. of about 1.5 V. A **battery** is made by joining two or more cells together, often to provide a larger e.m.f. (say 9 V).

Types of cells

Cells are described by the materials used in their electrodes and/or their electrolytes. There are two main groups of cells: **primary cells** which **cannot be recharged** and **secondary cells** which **can be recharged**.

Primary cells

This table shows the properties of some types of primary cells:

type	examples	e.m.f.	properties	typical use
zinc-carbon	HP2; SP2	1.5 V	inexpensive; short life	occasional use, e.g. radio
alkaline	Duracell	1.5 V	more expensive; longer life	constant use, e.g. clock
mercury	button cell	1.2 V	long life; small current	constant use, e.g. watches

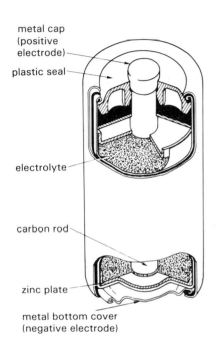

metal cap (positive electrode)

plastic seal

electrolyte

carbon rod

zinc plate

metal bottom cover (negative electrode)

Figure 51.1 Section through a typical dry chemical cell

From left to right: C, D, AA, PP3 and PP9 cells and batteries.

Secondary cells

This table shows the properties of some secondary cells:

type	examples	e.m.f	properties	typical use
nickel-cadmium	Nicad	1.25 V	medium current	stand-by supply, e.g. alarms
lead-acid	car battery	2 V	high current; heavy	motor vehicles (6 cells = 12 V)

Sizes of cells

In general the larger cells can provide more current for a longer time. The e.m.f. does not change with the size of the cell. Manufacturers have agreed on certain standard sizes.

Capacity of cells

The **capacity** of a cell indicates the total amount of energy it can supply before it 'goes flat'. Capacity is measured in Ah or mAh (amp-hours or milliamp-hours), so a 3 Ah cell can provide 0.5 A for 6 hours or 1 A for 3 hours, etc.

Current from a cell

Cells cannot provide a very large current because of their **internal resistance** (see chapter 50). So we could not expect a small 'C' size 3 Ah cell to provide 12 A (not even for ¼ hour). The only exception to this is the lead-acid cell which can provide very large currents.

Cells in series and parallel *(Figure 51.2)*

If two cells are in **series** their e.m.f.s add together but the current they can supply stays the same.

If two cells are in **parallel** the e.m.f. stays the same but they can supply more current.

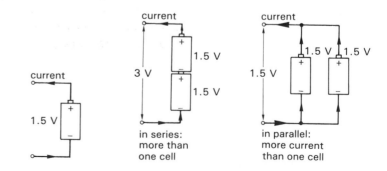

Figure 51.2 Combining cells. In series: more e.m.f. than one cell. In parallel: more current than one cell.

Electric wheelchairs provide mobility for the infirm. Its cells can be recharged overnight.

Rechargeable (secondary) cells

Rechargeable cells like the nickel-cadmium cell are a popular way of supplying portable power without having to buy new cells all the time. To get the best life out of them we must take some care:

● The cell must be discharged before trying to recharge it.
● They need special rechargers which provide a constant current. This adds to the cost.

Summary

● Cells are a portable way of supplying power.
● Primary cells cannot be recharged and are expensive to use.
● Secondary cells can be recharged but they cost more than primary cells and need a special recharger.
● Cells cannot provide much current and so cannot produce very much power.
● Lead-acid cells can provide much more power but they are heavy, bulky and expensive.

Questions

1   What is the difference between primary and secondary cells?
2   What is the e.m.f. of a battery of six 1.5 V cells?
3   Why would cells be no use for a 100 W amplifier?
4   What type of cell is best for:
    (a) a hearing aid   (b) a portable CB radio   (c) a clock?
5   A particular battery-powered digital stop-clock can measure up to 9.99 seconds in hundredths of a second. It works by counting pulses from an astable and displays the result on 7-segment LED displays.
    (a) How many 7-segment displays does it use?
    (b) If each segment of a display takes 5 mA, what is the maximum total current that could be drawn by the displays if they were all on constantly?
    (c) What is the average current drawn by the 'tenths' display during one second (i.e. as it cycles from '0' to '9')?
    (d) In practice, the supply to the displays is pulsed rapidly so that only one 7-segment display is on at any given instant. The pulsing frequency is so high that it gives the impression of a steady display.   (i) What is the maximum current drawn by the whole display now?   (ii) What effect might this have on the life of the battery?

52
Mains-driven Power Supplies

The **mains** is a useful source of power for a system. It can provide more current (therefore more power) than a cell. It is also much cheaper to use the mains. Unfortunately, the mains cannot be used as it is. The high e.m.f. (240 V) must be reduced to the low e.m.f. needed by most systems, and the a.c. must be turned into a steady d.c. A **mains-driven power supply** does this.

Transformers

A **transformer** can take the 240 V a.c. from the mains and turn it into a lower e.m.f. Figure 52.1 shows a typical transformer and its symbol. See chapter 73 for more details.

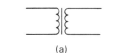

(a)

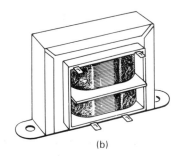

(b)

Figure 52.1 Transformers (a) Symbol (b) Example of a typical device

Using a transformer *(Figure 52.2)*

The primary is used as an input and connects to the mains. The secondary is used as an output. The e.m.f. which appears at the secondary depends on the number of turns in each coil. A common value is 12 V. Notice that it is still a.c.

Power rating (VA rating)

A transformer is **rated** according to how much current it can supply at its output voltage without overheating and failing. It is measured in VA (volt–amps) so a 12 V transformer rated at 6 VA can supply ½ A of current (12 × ½ = 6).

Rectifiers

A **rectifier** turns the a.c. from the transformer into a d.c. It is made from **diodes** which only allow current to flow through them one way. Figure 52.3 shows some typical diodes.

The simplest rectifier uses one diode. This simply cuts off the current when (for half of its cycle) the a.c. is the 'wrong way'. This process is called **half-wave rectification** (Figure 52.4).

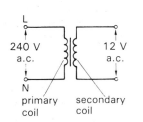

Figure 52.2 Transforming the 240 V a.c. mains into 12 V a.c.

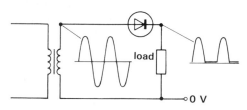

Figure 52.4 Half-wave rectification using one diode. Note that 'load' in this and in the following diagrams stands for the circuit which the power supply is connected to.

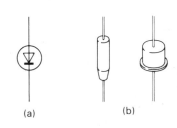

(a)   (b)

Figure 52.3 Diodes (a) Symbol (b) Typical examples

Full-wave rectifiers

Although half-wave rectification is simple, it wastes half of the transformer's output. It is also difficult to **smooth** (see later).

A **full-wave rectifier** turns the negative half of the a.c. cycle **upside-down** so that now the current flows the same way all the time. (See Figure 52.5.) At any one time in the a.c. cycle, the current will be going the right way for two of the diodes (either D_1 and D_3 or D_2 and D_4).

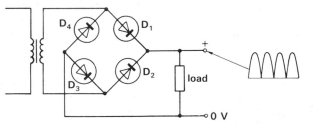

Figure 52.5 Full-wave rectification using a diode bridge

Bridge rectifiers

The arrangement of diodes in Figure 52.5 is sometimes called a **bridge rectifier**. Not surprisingly a bridge rectifier can be bought as a single package. Figure 52.6 shows one example.

Diode ratings

Diodes and bridge rectifiers are **rated** according to how much current they can stand before overheating and also the largest voltage that can be connected the **wrong way** before they conduct. A typical rectifier diode might be rated at 2 A and 100 V.

Centre-tapped transformers

A **centre-tapped transformer** has two secondary coils wound in opposite directions and joined at one end. This means that each secondary produces the same e.m.f. but their a.c. cycles are always **out of step**. Only two diodes are needed to produce a full-wave rectified d.c. from this. Figure 52.7 shows how this is done.

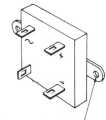

hole for mounting bolt

Figure 52.6 A typical bridge rectifier package

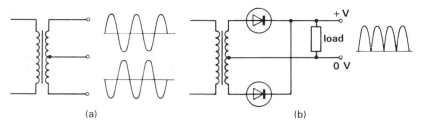

(a)   (b)

Figure 52.7 Centre-tapped transformers   (a) The output e.m.f.s from the transformer   (b) Full-wave rectification using two diodes

139

A typical smoothing capacitor. What do the markings tell you?

Smoothing

Although the 240 V mains has been transformed and rectified into 12 V d.c. it is not steady. The waves must be **smoothed** to give a pure, steady e.m.f.

This is done using a **smoothing capacitor**. The capacitor charges up when the voltage is high. When the voltage starts to go down, the capacitor gives up some of its charge. This keeps the voltage up and smooths out the wave. See Figure 52.8.

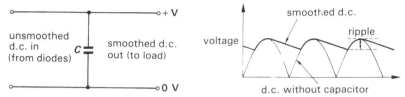

Figure 52.8 Smoothing a full-wave rectified d.c. with a capacitor. The ripple should be as small as possible.

More about smoothing capacitors

A smoothing capacitor must have a large value if it is to be effective. A typical value is about 2000 μF. This means that an **electrolytic capacitor** must be used. These are described more fully in chapter 72. Capacitors are also **rated** according to the largest voltage they can stand. A typical value might be 24 V.

Summary

● A transformer can take the 240 V a.c. mains and turn it into a lower voltage a.c.
● A rectifier turns a.c. into d.c.
● A capacitor can be used to smooth the d.c. produced.
● Each component has a **rating** which limits the voltage and current which can be supplied safely.

Questions

1 Draw a circuit for a power supply which takes the 240 V mains and produces a steady d.c. Explain what each component does.
2 Draw a diagram to show the differences between full- and half-wave rectification.
3 Why is full-wave rectification used most often?
4 What limits the amount of current that can be taken from a mains-operated power supply?
5 Some items of equipment use large capacitors in the power supply.
   (a) Why is it possible to get a shock from such equipment even when it has been unplugged from the mains?
   (b) Such equipment often has a resistor connected across each capacitor. What is the reason for this?
   (c) The value of this resistor is important. (i) What would happen if it were too small? (ii) What would happen if it were too large?

53
Safety

Although the transformer takes its energy from the mains there is no direct electrical connection between the secondary and the primary. If a low-voltage, mains-driven power supply is operated within its limits it is perfectly safe. Nevertheless accidents can happen so it is best to assume that they **will** happen and make sure that this will not damage people or the system.

Safety with the mains

We must be sure that not only are we safe from the dangers of the 240 V mains but also that the mains supply is not damaged if something goes wrong inside our power supply. Look at Figure 53.1.

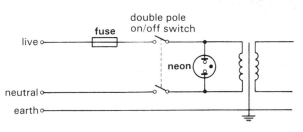

Figure 53.1   Safety features on the mains side of a power supply

On/off switch

A double-pole switch is best. This disconnects both live and neutral at the same time. If a single-pole switch is used it must be connected in the live wire.

'Mains on' indicator

A neon indicator glows when the mains is on. It takes very little current. It is also a better indication that the mains is on than a bulb powered from the low voltage side.

The unfortunate owner of this equipment did not bother to fit the correct fuse. All was well until a fault developed.

A typical power supply for a system. Try to identify each of the components and the safety features.

Fuse

The fuse is really a thin piece of wire. It normally does nothing. If the current should increase (perhaps because of a short circuit) the fuse will overheat, melt and stop any more current from flowing.

Fuse rating

Fuses are rated at the current they can normally stand. Typical values are 100 mA, 200 mA, 1 A and so on. If a power supply normally takes 150 mA from the mains then a 200 mA fuse should be used. A 1 A fuse would not blow until it is too late. A fuse is the only component where we always use the **smallest** rating that will work!

Earth

All metal parts which are not part of the actual circuit are connected to the earth wire in the mains lead. Then, if a live wire should touch the metal case, the current will flow straight down to earth. This current is usually enough to 'blow the fuse' or trigger a circuit breaker.

Case

Lastly, the power supply is put in a case and transformers, etc. are bolted down. This could be the same case that holds the rest of the system. If it is metal then the case must be connected to earth. This not only makes it safe, it also acts as a 'screen', helping to prevent interference.

Summary

- A mains-operated power supply must have an on/off switch connected in the **live** wire.
- A neon can be used to show when the mains is on.
- A fuse protects the circuits from damage when too much current flows. Fuses must be the correct type.
- All metal parts (including the case) must be earthed.

Questions

1 Describe how each of the following help to make a mains-operated power supply safe:
   (a) on/off switch   (b) a neon   (c) a fuse   (d) an earth wire.
2 Why is a bulb operated from the low voltage side not as safe as a neon?
3 How could you 'earth' a transformer?
4 If we use a single-pole on/off switch, why should it always be connected in the live wire?
5 Explain carefully what you would do if you discovered someone slumped unconscious over a piece of mains-powered equipment which they had been working on.

54
Voltage Regulators

We have seen that internal resistance in a power supply means that as more current is taken from the supply, the e.m.f. available to the rest of the circuits becomes less. Also, slight changes in the e.m.f. of the mains will affect the output p.d. of the power supply. These effects can be reduced by a **voltage regulator**. Its job is to keep the p.d. of the supply constant.

Zener diodes *(Figure 54.1)*

Zener diodes are described more fully in chapter 75. For now, we can think of them as a special type of diode which conducts a current when connected the 'wrong way round' to a p.d. of more than a few volts. The exact p.d. (called the Zener voltage) depends on the diode. Unlike resistors, a change in this current does not cause a change in p.d. across the Zener diode. Let us suppose we have a 9 V Zener diode and connect it to the rails of a power supply, as shown in Figure 54.2.

The p.d. of the power supply rails must be more than the Zener voltage of the Zener diode. The Zener diode has (in this circuit) 9 V across it. The remainder of the p.d. is taken by the resistor. Now, if the p.d. of the supply rails happens to decrease slightly then the resistor simply takes less p.d. The p.d. across the Zener diode stays constant. This is a simple **voltage stabiliser**.

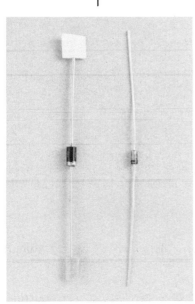

Figure 54.1 Zener diodes: symbol and examples of typical devices.

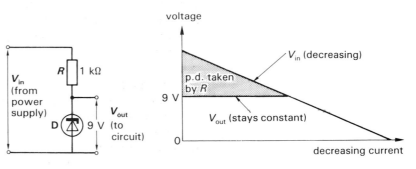

Figure 54.2 A simple voltage stabiliser

Voltage regulator

Unfortunately this simple arrangement cannot supply much current before the p.d. drops below the Zener voltage. If we take the supply current through a transistor, using the Zener diode to keep the voltage at the base constant, we have a more useful **voltage regulator**. (See Figure 54.3.)

Improved voltage regulator

An even better voltage regulator is shown in Figure 54.4.

The Zener diode is used to supply a **reference voltage** for a comparator. The comparator compares this reference voltage with a sample of the actual output voltage and controls a transistor which keeps the output voltage constant.

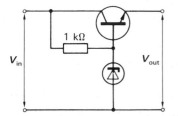

Figure 54.3 A simple voltage regulator. The transistor is normally a power transistor such as a BFY51.

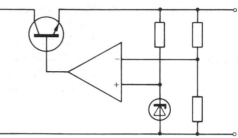

Figure 54.4 An improved voltage regulator

Integrated circuit voltage regulators

It is no surprise that such a useful circuit is available on a chip. It is very easy to connect into a power supply and for most purposes only a couple of capacitors (which act as **filters**) need to be added. A typical circuit is shown in Figure 54.5.

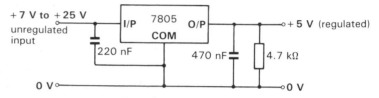

Figure 54.5 Voltage regulator using a 7805 IC and giving a steady 5 V output e.m.f. at currents up to about 1 A

Summary

- A Zener diode can be used with a resistor to make a simple voltage stabiliser.
- The amount of current allowed can be increased by using a transistor.
- A full voltage regulator uses the Zener diode to provide a reference voltage for a comparator which then controls a transistor.
- IC voltage regulators require very few extra components.

Questions

1. Why must the output voltage of a power supply be regulated?
2. Draw a simple voltage stabiliser and explain what happens if the p.d. of the power supply increases slightly.
3. Look through some catalogues and find an IC voltage regulator that can provide a regulated $+12\,V$ supply with a maximum current of 1 A. Draw a circuit which uses it.
4. A student made the circuit shown in Figure 54.6 to give 5 V from a 9 V supply.
   (a) Which component keeps the output voltage at 5 V?
   (b) What is the p.d. across R_1?
   (c) The LED should draw about 20 mA with a p.d. of 2 V across it. Calculate the p.d. across R_2 and hence suggest a value for R_2.
   (d) The circuit was correctly built, but the LED did not light. For each component, suggest a possible fault which could cause this malfunction.

An integrated circuit voltage regulator. Find out what voltage this one provides.

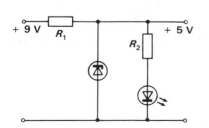

Figure 54.6

144

55
Exercises

1   The single cell shown in Figure 55.1(a) has an e.m.f. of 1.5 V.
(a) What is the e.m.f. of each of the combinations shown?
(b) What is the advantage of combination (b) over the single cell shown in (a)?
(c) Describe one advantage and one disadvantage of using cells as a power supply instead of using the e.m.f. of the mains.

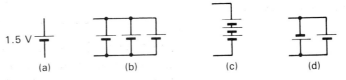

1.5 V

(a)   (b)   (c)   (d)

Figure 55.1

2   A circuit for a mains-driven power supply is shown in Figure 55.2.
(a) What differences are there between the e.m.f. produced by the mains and the e.m.f. produced at the output?
(b) There are three safety devices in the circuit shown. What are they and how do they help to make the system safe?
(c) What is the device labelled 'BR' called and what does it do?
(d) What type of capacitor is shown? What does it do?
(e) Draw a diagram to show what the output e.m.f. would look like when connected to an oscilloscope if: (i) capacitor C is the correct value   (II) capacitor C is too small   (iii) capacitor C is removed altogether.
(f) What will happen to the output e.m.f. if a fairly large current (but not enough to cause damage) flows from the output? Explain your answer.
(g) What circuit or device could be connected to the output which would prevent or reduce the effect in part (f) above? What is the name of the type of diode that this circuit or device uses?

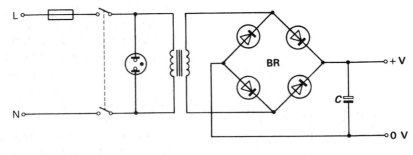

Figure 55.2

Section 7

Communication Systems

56 Communication

One of the greatest effects that electronics has had on our lives has been in the ways in which we communicate with each other, especially over long distances. Here are some examples.

Radio

The advantage of radio is that no direct connection is needed. This not only makes it simple to set up a communication link but it also allows the transmitter or receiver to move about.

A European geostationary satellite sends constant weather reports to stations on the ground. What else are satellites used for?

A driver trying to find his way in an unfamiliar town gets help over his CB radio. Who else uses two-way radio?

Telephones provide a vital communications link, especially in remote areas.

Television

We can often understand information more easily and quickly if it is in a visual form. In some cases pictures can be sent by radio waves. In other cases, a cable link is more satisfactory.

Telephone

Although it needs connecting wires the telephone system has advantages over radio. It can be more reliable and suffers from less interference. It is also more private.

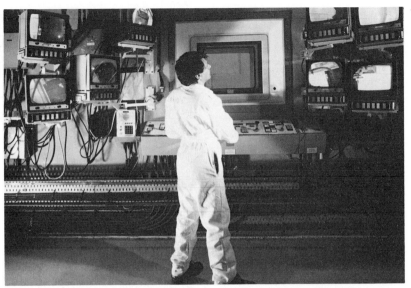

An engineer working with radioactive fuel elements. Using closed-circuit television, we can see inside installations which would be difficult or dangerous for people to enter.

Cellnet telephones combine the advantages of telephone and radio.

So far over 80 000 people have used their television to help them study for a degree through the Open University.

Peter has no strength in his arms. With
the help of electronics he can type using
a pointer on his head.

Other types of communication

The production of books and newspapers is helped by electronics,
especially computer systems. Electronics can help people with
disabilities to communicate more easily. Can you think of any
other types of communication which are helped by electronics?

This book was written on a computer word processor. The screen shows the text on
page 1.

Summary
- Radio, television and the telephone are all communication
  systems. They are only possible because of electronics.
- We use communications to keep people informed, to help in
  emergencies and simply to talk to each other.

Questions
1 What ways are there of communicating using electronics?
2 Make a list of the ways in which *you* use communication
   systems in everyday life.
3 Mountain rescue teams carry radio transmitter/receivers.
   (a) Explain two reasons why a rescue team might need to
   transmit as well as receive information.
   (b) Give three important considerations to be borne in mind
   when designing a two-way radio for a mountain rescue team to
   carry.
   (c) How can modern electronic components make it easier to
   design a portable two-way radio which meets these
   requirements?
   (d) Give two items of information that a rescue helicopter
   might send to the team on the ground.
   (e) What differences might there be between the radio in the
   helicopter and the radio carried by the ground team? What
   effect might these differences have on the range over which the
   radios can transmit?

57
Radio Waves

Any wire carrying an alternating current will give out (or **radiate**) electromagnetic waves. These waves will spread out, travelling at the speed of light.

If these waves hit another wire, an alternating current will be **induced** in that wire. The frequency of this a.c. will be the same as the original a.c. that produced the waves.

This is the basis of **radio** transmission and reception, discovered around 1880 by H R Hertz.

Radio frequency

It was soon found that low frequency waves did not travel very far before they became too weak to be picked up by the receiving wire (or **aerial**). Higher frequencies did much better. We call these **radio frequencies** or **r.f.**

Radio frequency starts at about 3 kHz and goes up to 300 GHz.

Wavelength

The **wavelength** of a radio wave can be found by dividing its **speed** (which is always 300 000 000 metres per second in air) by its **frequency**. The answer will be in metres.

The Greenlowther transmitter in Scotland. Equipment must be maintained, even under extreme conditions like these!

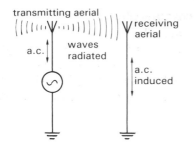

Figure 57.1 Transmitting and receiving radio waves

The radio spectrum

The full range of radio frequencies can be divided into various **bands** of frequencies. Six of these bands are shown here:

frequency	wavelength	name	old name
3 kHz–30 kHz	100 km–10 km	very low freq.	
30 kHz–300 kHz	10 km–1 km	low freq.	long wave
300 kHz–3 MHz	1 km–100 m	medium freq.	medium wave
3 MHz–30 MHz	100 m–10 m	high freq.	short wave
30 MHz–300 MHz	10 m–1 m	very high freq.	
300 MHz–3 GHz	1 m–10 cm	ultra high freq.	

Transmitting and receiving radio waves

An r.f. (i.e. radio frequency) **oscillator** is used to make the r.f. alternating current in a **transmitting aerial**. This then sends out r.f. waves which are picked up by the **receiving aerial** and turned back into an alternating current. (Figure 57.1.)

Radio signals

This simple process does not send any **information**, apart from telling the receiver that someone is sending out a radio wave! Just as an electrical signal is made by changing a steady voltage or current, a radio signal is made by changing the steady radio wave in some way. Usually the signal we want to send is either speech or music. Both these types of signals consist of fairly low frequency **audio signals** (a.f.). The frequency of the radio wave used (r.f.) is much higher. The r.f. signal is used as a **carrier** for the audio signal. There are several ways of doing this. We call it **modulating** the radio wave. Figure 57.2 shows a block diagram of how this is done.

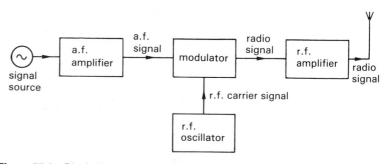

Figure 57.2 Block diagram of a radio transmitter

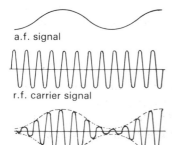

a.f. signal

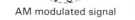

r.f. carrier signal

AM modulated signal

Figure 57.3 Amplitude modulation

Amplitude modulation

In **amplitude modulation** (or **AM**) the audio signal is used to control the **amplitude** (size) of the carrier radio wave. (Figure 57.3.)

The receiver can pick up the r.f. wave and then **detect** the changes in its amplitude to find the a.f. signal being transmitted.

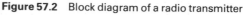

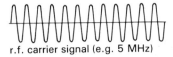

a.f. signal (e.g. 4 kHz)

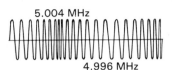

r.f. carrier signal (e.g. 5 MHz)

5.004 MHz

4.996 MHz

FM modulated signal

Figure 57.4   Frequency modulation

Advantages and disadvantages of AM

● AM receivers can be very simple to make.
● AM transmissions can suffer from **interference**. Many things can change the amplitude of a radio wave as it travels (e.g. atmospheric conditions). These will affect the signal which is finally detected. Everybody knows what some AM radio stations sound like in the evening!

Frequency modulation

We could vary the frequency of the carrier instead of the amplitude. Once again, the audio signal controls how much the carrier is varied (Figure 57.4). This is **frequency modulation** or **FM**.

Advantages and disadvantages of FM

● It is much harder to affect the frequency of a radio wave as it travels, so FM suffers from less interference.
● FM receivers are more difficult and expensive to make.
● They need much more 'space' in the frequency bands (i.e. they need more **bandwidth**). (See Figure 57.5.)

Figure 57.5   When we transmit by FM we must leave about 10 kHz between stations to allow room for the a.f. signals to be added to the FM carrier

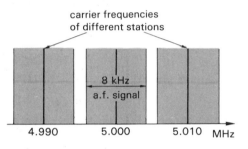

carrier frequencies
of different stations

8 kHz
a.f. signal

4.990   5.000   5.010   MHz

Summary

● Radio waves are transmitted by producing a high frequency alternating current in a transmitting aerial.
● These waves produce a similar a.c. in a receiving aerial.
● An r.f. wave is used as a carrier for the a.f. signal being sent.
● Amplitude modulation varies the amplitude of the carrier wave. It can suffer from interference.
● Frequency modulation varies the frequency of the carrier wave. It does not suffer from so much interference but it needs a large bandwidth.

Questions

1   What does a transmitting aerial and a receiving aerial do?
2   What do the following abbreviations stand for?
    (a) r.f.   (b) a.f.   (c) AM   (d) FM   (e) u.h.f.   (f) MHz
3   (a) What is the highest and the lowest radio frequency?
    (b) Why are such high frequencies used?
4   Explain with diagrams the difference between AM and FM.
5   Why is FM only used with carrier waves in the higher bands?

58
Simple Radio Receivers

A radio receiver must be able to pick up radio signals, sort out the station we want and then recover the audio signal being sent to finally produce sound in the loudspeaker. A block diagram of a simple radio receiver is shown in Figure 58.1.

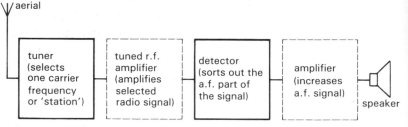

Figure 58.1 A tuned radio frequency (TRF) receiver. The dotted blocks can be omitted in very simple receivers.

Simple diode (or crystal set) receiver

This is the simplest circuit which can receive AM transmissions (Figure 58.2):

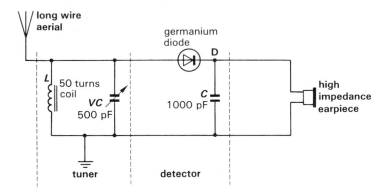

Figure 58.2 A crystal set receiver

The circuit has no amplifier. The earpiece simply turns the a.c. signals produced in the aerial straight into sound waves. Early receivers used a diode made from a thin piece of wire, called a 'cat's whisker', touching a 'crystal'.

Aerial and tuner

The **aerial** picks up all the incoming radio waves and turns these into alternating currents at different frequencies; one for each radio wave. This mixture of signals goes into the **tuner**.

The coil and capacitor (*VC*) together make a **tuned circuit** (see Figure 58.3). This lets all the signals flow straight down to earth, apart from the one with the frequency that we want. This signal is forced to go into the next stage, the **detector**.

By altering the variable capacitor we can 'tune in' to different frequency signals.

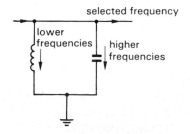

Figure 58.3 A simple tuned circuit

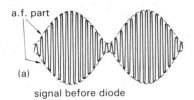

a.f. part

(a)

signal before diode

(b)

signal after diode (but without C)

(c)

signal with C in circuit

Figure 58.4   Detecting the a.f. part of the signal with a diode and capacitor

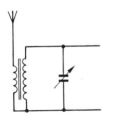

Figure 58.5   The selectivity can be improved by using two coils on the same ferrite rod

Detector

If we simply let the signal go straight into the earpiece we would not hear anything. The a.f. part of the signal is trying to make the earpiece go both ways at once and the r.f. part is too high to hear (Figure 58.4a).

The diode cuts off the negative part of the signal (Figure 58.4b). This now lets us **detect** the a.f. part of the signal.

The capacitor C filters out the high frequency (r.f.) part of the signal (Figure 58.4c) so that what finally goes into the earpiece is the a.f. signal which we can turn back into sound.

Faults

There are three main faults with this simple circuit:
1   It is not very **selective** (i.e. it is difficult to sort out stations which are close together on the frequency band).
2   It is not very **sensitive** (i.e. it cannot pick up weak or distant stations).
3   The sound produced is very small and of poor quality.

Improvements

● One way that we can improve the simple receiver is to **amplify** the a.f. signal before turning it into sound. A simple transistor amplifier (see chapter 41) gives good results. This does nothing towards improving the **selectivity**, however.
● The sensitivity and selectivity can be improved by changing the design of the coil. One way is to use two coils as a type of transformer. An example is shown in Figure 58.5. Even so, there is a limit to how good this type of receiver can be. The next chapter shows how a different design can produce much better results.

Summary

● A simple TRF (tuned radio frequency) receiver has five stages: aerial, tuner, detector, a.f. amplifier and speaker.
● A simple diode receiver uses:
   (a) a coil and capacitor for a tuner
   (b) a diode as a detector
   (c) an earpiece to produce sound.
● The simple TRF receiver is not very selective.
● The circuits in this chapter work only on AM transmissions.

Questions

1   What does TRF stand for?
2   Draw a block diagram of a simple TRF receiver and describe what each part does.
3   Draw diagrams to show what the waveform of the signal looks like before the detector and after the detector.
4   What are the three faults of a simple diode receiver?
5   Which of the faults you described in question 4 cannot be improved very much?

59
Improved Radio Receivers

We can overcome problems of poor selectivity of the TRF receiver described in the last chapter by adding some extra stages to the circuit. The result is a **superheterodyne receiver** (or **superhet**). To understand how this works we must first look more closely at waves.

Mixing signals *(Figure 59.1)*

If two signals with different frequencies are sent into a **signal mixer** they become mixed together.

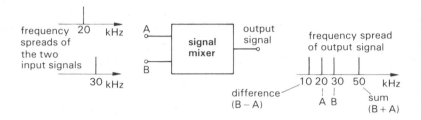

Figure 59.1 Mixing two signals with different frequencies

The signal coming out of the mixer contains **four** main parts; the two original signals and two **new** signals. One of them is the **sum** and the other is the **difference** between the two original signals.

The difference signal

Suppose that we send all this into a filter and remove three of the signals, leaving just the **difference signal**. This signal will have all of the characteristics of the two original signals (A and B). If one of them (say A) is a modulated signal then the difference signal will be modulated in exactly the same way.

If we now change the frequency of signal B then this will simply change the frequency of the difference signal.

In effect we can take a modulated signal (signal A) and change its frequency to whatever we like by this process, which is called **heterodyning**.

Superheterodyne receiver

The principle of frequency changing is used in the **superhet** (or **superheterodyne**) receiver. Figure 59.2 shows a block diagram:

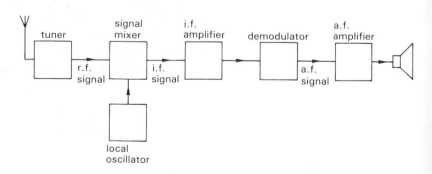

Figure 59.2 Block diagram of a superhet receiver

A ganged variable capacitor.

How it works

Signal A is the incoming radio signal. Signal B is produced by a **local oscillator**. The **difference signal** is called the **intermediate frequency** or **i.f.** signal.

The i.f. signal contains all the information carried by the original r.f. signal, but at a lower frequency (usually 465 kHz).

The i.f. amplifier is a **tuned amplifier** which only amplifies signals inside a small frequency range (around 465 kHz) and in effect 'tunes in' to the signal we want.

The signal is then **demodulated** to detect the original a.f. signal being transmitted.

To tune in to a different station we use a **ganged variable capacitor**. Half of it alters the **tuner** and the other half changes the frequency of the **local oscillator** so that a new radio signal becomes the i.f. signal.

Advantages of superhets

- It is better to design an amplifier to work on just one frequency (the i.f. amplifier) rather than trying to amplify every radio signal, especially since the i.f. is a lower frequency.
- The i.f. amplifier in effect adds another tuning stage by rejecting all signals outside the intermediate frequency.
- The result is a receiver which is more selective than the TRF described in the last chapter.

Summary

- A superhet receiver is more selective than a TRF receiver.
- It uses an oscillator and a mixer to change the frequency of a radio signal to a lower frequency (i.f.) signal.
- The i.f. amplifier only amplifies the i.f. signal.
- The signal is then demodulated.
- If we change the frequency of the local oscillator then a different radio signal will become the i.f. signal.

Questions

1 Why is a superhet receiver better than a TRF receiver?
2 What does each of the following do?
  (a) local oscillator (b) mixer (c) i.f. amplifier
3 What does 'i.f.' stand for? What is its usual frequency?
4 What signals would you get if you put a 2.2 MHz signal and a 2.8 MHz signal together in a mixer?
5 What frequency must the local oscillator have to produce an i.f. of 465 kHz from a radio signal at 1.2 MHz?

60
Television

Television uses radio waves to carry information which is finally turned into pictures on a **cathode ray screen**. This information, and the system needed to produce and process it, is quite complex, but like all systems it can be broken down into smaller building blocks.

How a TV signal is produced: TV cameras

Light (from the image being filmed) enters the camera and is focused on a **detector**. The detector consists of thousands of light sensors (LDR and resistor) arranged in a regular pattern of 625 separate rows.

The detector screen is **scanned** one row at a time under the control of electronic switches. Each scan produces an electrical signal which depends on how much light was falling on each sensor in that row (Figure 60.1). The next line is then scanned and so on (625 lines in all) until the whole **frame** has been scanned.

The whole process is repeated 25 times per second.

A **colour camera** has three detectors, one each for red, green and blue light from the image.

The information (together with **synchronisation pulses**, etc.) is then transmitted by amplitude modulation on UHF radio wave. At the same time a sound signal is broadcast by FM.

TV screen

A TV screen is very similar to an oscilloscope screen. **Phosphors** on the inside of the glass at the front 'light up' when hit by a beam of electrons from inside the tube (Figure 60.2).

A black and white TV uses just one type of phosphor, which glows white. A colour screen has three other types of phosphors, arranged in a regular pattern. One set glows green, another glows red and the third glows blue. Three electron beams are used to **excite** each phosphor as needed.

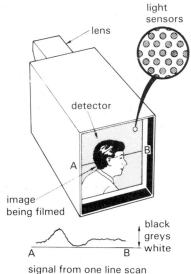

Figure 60.1 A TV camera produces a signal by scanning the image a line at a time

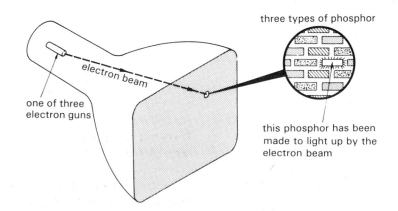

Figure 60.2 Simplified diagram of a TV tube showing one phosphor which has been made to glow by the electron beam. The path of the electron beam is controlled by electromagnetic coils which are not shown here.

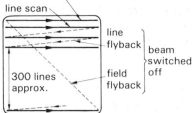

1st line of next field
line scan
line flyback
beam switched off
300 lines approx.
field flyback

Figure 60.3 The raster pattern of scanning on a TV screen

How a picture is produced

This is like the TV camera in reverse. The electron beam scans the phosphors in a pattern called a **raster**, turning on and off as it goes. This produces a pattern of light and dark on the screen. It scans 312 lines to fill the **field** and then goes back to fill in the alternate lines. This is called **interlacing** (Figure 60.3).

A complete picture (or frame) is made up from these two fields, 625 lines in all; 25 frames are produced per second.

If we use the signal from our camera to turn the beam on and off we will end up with our original image.

Field blanking

Although 625 lines are transmitted, the first and last 18 or so lines in each field are 'blanked out' and are not shown on the screen. This is to allow time for the beam to move back to the top of the screen and let the circuits settle ready for the next scan.

The actual picture really consists of about 550 lines in all.

The TV receiver

A block diagram of the main parts of a colour TV receiver is shown in Figure 60.4:

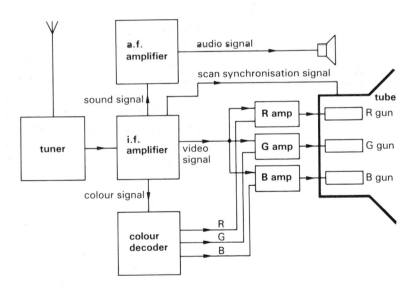

Figure 60.4 Simplified block diagram of a colour TV receiver

How it works

After the **tuner** (which works on the superhet principle) the signal is amplified by the i.f. amplifier and split into four separate signals.

Each electron beam is controlled by two signals: the video signal and the colour signal.

The **video** signal is used to control all three electron beams together and so produce the pattern of light and dark (red + green + blue light = white light!).

157

The **colour** signal is split into three (R, G, B) and gives extra control over each electron beam separately, 'painting in' the colour over the basic black and white picture.

The **scan synchronisation** signal is used to control the circuits which make the three beams scan the tube. This is very important. It makes sure that each beam carries out its scan and flies back at exactly the right time.

The **audio signal** is used to produce the sound.

Summary

- A TV camera scans the light falling on a detector to produce a signal.
- Each picture is called a frame and consists of two fields. The whole frame contains 625 lines.
- The TV picture is transmitted using AM on a UHF carrier wave.
- A colour TV screen has three types of phosphors (red, green and blue) which are lit up when their electron beam strikes.
- Circuits inside the TV set scan the electron beams across the tube in a raster pattern over 625 lines.
- The TV signal is used to control the 'brightness' of each beam as it scans, so producing a picture.
- It is important that the whole process is synchronised.

Questions

1 What is meant by 'scanning'?
2 What is meant by a 'field'? How many lines are in a field?
3 What is a 'frame' and how many lines does it have?
4 A TV signal has to put a lot of information into a short time. Why is the signal transmitted using AM and not FM?
5 What is 'synchronisation'? What do you think a TV picture might look like if there was no synchronisation?
6 Much research is currently being done into designing a different type of television screen. This will have many advantages, some of which are listed below. For each one, explain how it might affect both the design and the uses of television sets.
    (a) Less voltage and current will be needed to drive the screen.
    (b) The screen could be made extremely thin and light.
    (c) The screen would last much longer.
    (d) The screen would be easier to mass produce.

61
Teletext

Teletext is an example of a system which combines digital circuits and the television system. It allows users with a suitably modified TV set to receive up-to-date information on a variety of subjects. The information is transmitted by the TV companies along with their normal broadcasts. The BBC version is called **Ceefax**, while the IBA call theirs **Oracle**.

The teletext display

Each **page** of teletext consists of **24 rows** of information. Each row can contain up to 40 simple **characters**.

An example of a teletext page is shown in Figure 61.1.

Transmitting teletext signals

We saw in the last chapter that only about 550 of the 625 lines are used for the actual TV picture. Four of these 'unused' lines (17, 18, 330 and 331) are used to transmit the digital teletext signals. (See Figure 61.2.)

Figure 61.1 A typical teletext display. The pictures are very simple because only 24 rows are used.

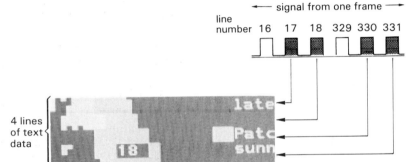

Figure 61.2 Four lines in every frame signal each carry a row of teletext data

Each line carries the signals for one **row** of a teletext display. So after 6 frames a complete page of 24 rows has been transmitted. The next page is then transmitted and so on. Since 25 frames are transmitted each second, we can transmit about 4 pages each second. In all several hundred pages might be transmitted before starting again at page 1.

These signals are **decoded** by the TV set which can display the teletext page instead of the normal picture.

62
The Telephone System

It is quite simple to arrange a system to let one person talk to another over a distance; a microphone and earpiece linked by a pair of wires will do. What makes the telephone system so important is the network which allows every telephone to be connected to any of the other telephones in the system. The telephone **exchanges** make this possible.

Telephone exchanges

Telephones are linked to a nearby **local exchange** (Figure 62.1). When one telephone dials another local telephone, a network of switches in the exchange connects the wires from the two telephones together.

The exchanges are themselves linked together (often through larger **area exchanges**) so that two telephones in different areas can be connected together (Figure 62.2).

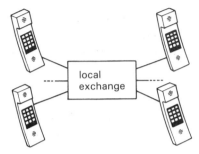

Figure 62.1 Telephones are linked to a local exchange

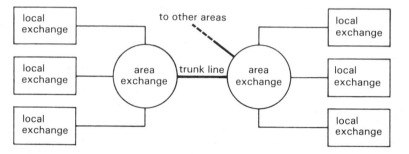

Figure 62.2 Exchanges are linked by trunk lines

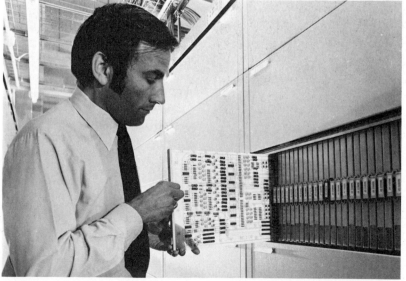

(*Left*) Inside the Newmarket telephone exchange, built in 1949. The banks of relays make the connections for the calls. (*Right*) Inside the first System X exchange. This can handle a similar number of calls but faster, more reliably and with better quality. Why might the telephone engineers prefer the System X exchange?

There are thousands of switches in each exchange. When you dial a number, your phone sends a series of pulses to the exchange. The exchange uses these pulses to control the switches which make the connection you want.

At one time all the exchanges used electromagnetic relays as switches. Now these are being replaced by logic gates and latches. Because there are no moving parts, the new exchanges are more reliable and can work more quickly than the old types.

System X

Since 1984 British Telecom have been steadily replacing most of the old telephone system with the all-electronic System X. Digital electronics are used throughout. The new exchanges are just part of this system.

Computer data links

Not only telephone conversations are sent through the system; computers can exchange data along telephone lines using **modems**. (See chapter 65.)

Information services

Information can be held in a computer **database** and then sent through the telephone system to other computers.

Figure 62.3 is a page from **Prestel**. It is a public information service, similar in some ways to teletext. Users can dial the service, and the data they want is sent along the telephone system into their own computer where it can be displayed on a VDU.

There are other databases linked to the telephone system. One important database, for example, contains details of court cases and judges' decisions. This gives lawyers instant access to a vast amount of information which would otherwise fill hundreds of books and soon be out-of-date.

The London Stock Exchange uses computers extensively for information processing. The telephone network is used as a data link to connect them to other computers throughout the world.

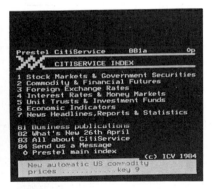

Figure 62.3 A page from Prestel. The database holds over 100 000 pages of information which are constantly updated.

Questions

1 What were the main uses of the telephone system? What are they now?
2 Look at the two photographs on page 160. Which system do you think Telecom engineers prefer? Why?
3 System X uses digital signals wherever possible. Why?
4 British Telecom could have started changing to a digital system years ago if they had bought equipment available in other countries. Instead they waited for British firms to develop their equipment. Was this a good idea?

63

Exercises

1 Shown in Figure 63.1 is a block diagram of a radio transmitter. Throughout the system, information is being changed from one form to another (e.g. sound waves, electrical signals, etc.).

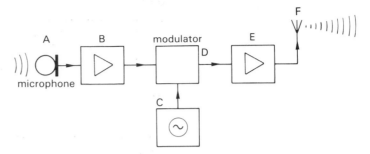

Figure 63.1

(a) For each of the blocks A to F state: (i) What type of information or signal goes in to the block. (ii) What type of information or signal goes out from the block. (iii) The name of the device or circuit in the unlabelled blocks.

(b) State a typical frequency for the signal going out of block B.

(c) State a typical frequency for the signal going out of block C.

2 Two types of modulator are represented by these blocks (Figure 63.2):

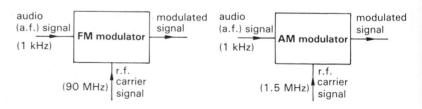

Figure 63.2

a.f. signal

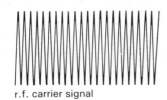

r.f. carrier signal

Figure 63.3

(a) What do FM and AM stand for?

(b) Why is the carrier signal much higher frequency than the audio signal?

(c) Suppose the signals shown in Figure 63.3 were sent into the modulators. (i) Draw a diagram to show what type of signal would come out of each modulator. (ii) Describe one advantage and one disadvantage of FM. (iii) Describe one advantage and one disadvantage of AM.

3   Shown in Figure 63.4 is a block diagram of a simple TRF receiver:

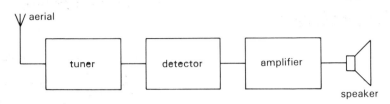

Figure 63.4

(a) Explain briefly what comes in and what goes out of each block.
(b) Draw the simplest circuit or device that could be used for each block.
(c) Describe the limitations of the simple TRF receiver.
(d) What additional blocks are needed to make a superhet receiver? Describe briefly how this improves selectivity.

4   Radio has many uses. Some examples are:
(1) in private communication between two people
(2) in mass communication of information to many people
(3) in warfare
(4) communication with satellites and spacecraft
(5) sending instructions to other devices (e.g. model boats).
Choose one of these and describe:
(a) The advantages that radio has over other methods.
(b) Any problems that might arise from using radio rather than another method.
(c) Ways in which these problems might be overcome.
In your answers, consider aspects such as **security, inter-ference, corruption of data, complexity of equipment, cost** and so on.

5   (a) Which frequency band is used for transmitting television?
(b) What method of modulation is used for the video signal?
(c) How many complete pictures are transmitted each second?
(d) How many lines of information are transmitted for each picture?
(e) How many lines are actually shown on the TV screen?
(f) What are the missing lines used for?

6   (a) What kinds of information are transmitted as teletext?
(b) Are teletext signals digital or analogue?
(c) Explain briefly what a teletext page looks like on a screen.
(d) Where, in the TV signal, is the teletext signal transmitted?

Section 8

Programmable Systems

64 Computer Systems

Computers are very much part of our everyday lives. They can store large amounts of information, carry out complicated calculations very quickly, control processes and so on. Although a computer system may seem impossibly complicated it can still be broken down into simpler building blocks just like any other electronic system. A block diagram of a typical computer system is shown in Figure 64.1.

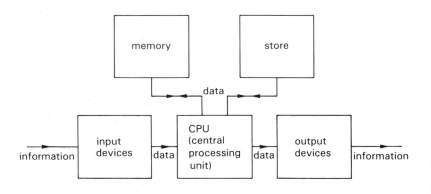

Figure 64.1 Block diagram of a computer system

These notes will help in understanding the block diagram:

- **Information** may be in a form that humans can understand (e.g. letters, numbers, sound waves, etc.) or it may be electrical signals from **sensors** (e.g. a heat sensor).
- The **input device** turns information into digital signals (**data**).
- The **central processing unit** (CPU) works to a list of instructions called a **program**. The CPU takes the data and either stores it for later use or **processes** it (perhaps doing a calculation) and sends it to the output.
- The **output device** turns the digital signals back into information. Once again this may be in a form that we can understand or it may be signals to another device or system.

The store

There is a limit to how much can be held in the memory. The **store** is used for data that is not needed at present. There are several ways of storing data. Two common methods both use tape heads to store digital signals as patterns of magnetism.

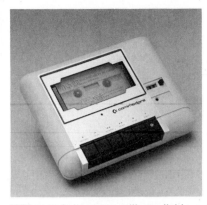

Without a low-cost, readily available storage device like this cassette tape recorder, it is doubtful whether home computers would have become as popular as they are.

One floppy disc of the type being used here was used to store all the text for this book while it was being written.

Storage devices

Disc drives use a magnetic disc (floppy disc or hard disc). The tape head can move to any point on the disc so that any piece of data can be reached quickly.

The cassette recorder is used by many home computers. The tape may have to be wound through before the data required can be reached. This is much slower than disc.

Summary
- A computer system contains a central processing unit, a store and at least one input and one output device.
- An input device turns information into digital signals.
- An output device turns data back into information.
- The CPU processes the data under instructions from the program.
- Data can be held in a store. Two common ways of storing data are magnetic disc and tape.

Questions
1   Explain, using a diagram, what is in 'a computer system'.
2   What is meant by 'data'?
3   Magnetic tape and magnetic discs are two ways of storing data. Which one is faster and why?

65
Input and Output Devices

Input devices
At least one input device is needed; many computers have more.

Keyboard
This is a system of switches. They may be mechanical **keys** or electronic switches (e.g. touch sensitive). Some electronics are needed to make a different digital signal for each switch that is pressed.

Light pen
This is a light-sensitive switch in a holder. It produces a simple signal each time it receives light from the VDU screen. The CPU controls how and when each part of the screen is lit up so it can work out where the light pen is when it sends in its data.

A light pen is extremely useful in computer-aided design (CAD) and drawing programs.

Magnetic reader
The magnetic strip on the back of a plastic cashcard is read by a **magnetic tape reader**. It turns previously recorded patterns of magnetism on the card into digital signals.

Other input devices
Computers which are in control of machinery and equipment use various sensors to keep a check on things like **temperature**, **acidity**, **position of objects** and so on. Each of these sensors works in a different way and may include an **analogue-to-digital converter** (see chapter 27), but they eventually produce data.

Output devices:

VDU
This stands for **visual display unit**. Although it uses a screen like a TV, the picture is produced from digital signals rather than a modulated wave. Some home computers contain a **TV modulator** which turns digital signals into a signal like a TV signal so that a TV set can be used instead, but the screen display is not as good. A TV screen is not designed to display the same amount of detail as a VDU.

A computer compares the data read by the magnetic reader from the card with data entered by the customer on the keys. If the two sets of data are the same, the computer will accept the customer's instructions.

Printers and plotters
There are several ways of producing print and drawings:

A **daisy wheel printer** is a bit like a typewriter (without a keyboard) controlled by signals from the computer.

A **dot matrix printer** is faster. It makes up each character from a pattern of dots. The print is not as good quality as from a daisy wheel.

A **plotter** draws lines rather than prints characters. It can be used to produce drawings and diagrams.

Part of a daisy wheel. You can see how it got its name.

Other output devices

Speech synthesisers use digital signals to build up a pattern of sound like a human voice.

Robot arms use data to control their movements.

A **modem** (or modulator/demodulator) is really two devices, an input and an output device, in one box. It is used to allow two computers to send data to each other over the telephone system. The **modulator** part turns digital data into audio signals which are sent to the other computer. This uses its **demodulator** to turn these signals back into digital data.

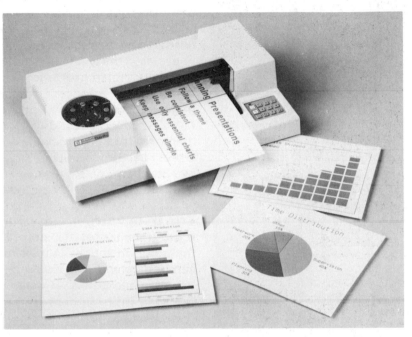
A plotter can produce high-quality drawings on overhead transparencies as well as on paper.

Summary

● An input device turns information into digital signals (data). Two input devices are a keyboard and a light pen.

● An output device turns data back into information. Two output devices are a VDU and a printer.

● Computers use a modem to convert data into signals which can be sent through the telephone system to each other.

Questions

1   Which input and output devices are best for:
    (a)   working out a payroll and printing out payslips
    (b)   designing a new car and producing the plans
    (c)   sending messages to another firm's computer?

167

66

The Processor

The central processing unit (CPU) is the heart of the computer. This is where the data is processed.

Inside the CPU

The CPU has three main parts (Figure 66.1):

The **arithmetic/logic unit (ALU)** is a set of logic gates. It can carry out simple calculations or compare two numbers to see if they are different.

The **registers** are flip-flops which temporarily hold data during calculations.

The **control** looks after the process and handles the input and output of data.

Microprocessor

These days the whole of the CPU is in an integrated circuit. This is known as a **microprocessor**.

Instructions

The CPU collects its instructions, one at a time, as data held in the memory. These instructions are very simple such as 'fetch some more data from the input or memory', 'do a calculation', 'compare two numbers', or 'send data to the memory, output or store'. Then it must ask the memory for the next instruction.

The program

A **program** is a list of instructions. Although it is possible to write the program entirely in the simple instructions that the processor can understand, this would be very tiresome. Figure 66.2 shows the instructions needed just to add two numbers.

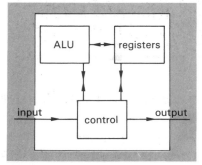

Figure 66.1 General structure of the Central Processing Unit (CPU)

The Z80 CPU is used in many computers. The actual silicon chip is about 1 mm square and is connected by very fine wires to 40 leads which go through the side of the case.

instruction	meaning
0 1 1 1 1 1 1 0	Put into register A . . .
0 0 1 0 1 0 0 0 0 1 1 0 0 1 1 1	the number held . . . in this memory location.
0 1 0 0 0 1 1 0	Put into register B . . .
0 0 0 1 0 1 1 0 0 0 1 0 0 0 1 1	the number held . . . in this memory location.
1 0 0 0 0 0 0 0	Add them together.
0 1 1 1 0 1 1 1	Put the result . . .
0 0 0 1 0 0 1 0 0 1 1 1 0 0 1 1	into this . . . memory location.

Figure 66.2 A typical list of instructions needed by a Z80 CPU to add two numbers together

Computer languages

In general we write programs in a **high-level computer language** (e.g. **Basic**). This makes it easier for us. The computer then translates this into the simple codes that the CPU can understand. The computer has a 'dictionary' stored in its memory.

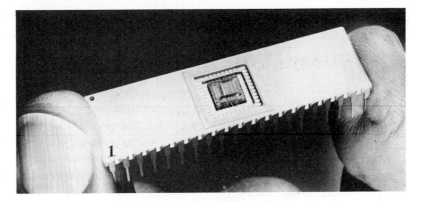

The window in this integrated circuit is normally covered by a small, metal plate. When the plate is removed the circuit can be exposed to ultra-violet light.

The memory

There are three main types of data which must be stored in the memory:

- The **list of instructions** which make up the program.
- The **data being processed**, i.e. the text or numbers which have come from the input device in the form of data.
- The **dictionary** or **interpreter** which the computer uses to translate the program language into the language of the CPU.

ROMs and RAMs

Some of the data may change while the computer is working; perhaps new data may come in or we may alter the program. This type of data is held in a **random access memory** (**RAM**).

The dictionary, however, will not change. It is held in a **read-only memory** (**ROM**).

The RAM consists of thousands of bistables. Each one can store a logic 1 or logic 0. Of course, when the power is turned off all the data stored in the RAM is lost.

The ROM also contains a type of bistable. These have been **fused** by the manufacturer so that, once they have been set, they can never be changed again. Data held in the ROM stays there, even when the computer is turned off.

An EPROM is an **erasable programmable read-only memory**. It behaves like a ROM but has the advantage that it can be 'wiped clean' (usually by shining ultraviolet light on it) and then re-programmed by a special device.

Each of these types of memory is contained in ICs.

Summary

- The CPU can carry out only simple instructions.
- A program is a list of instructions. High-level languages like Basic must be **interpreted** by the computer using a kind of **dictionary** held in the ROM.
- There are two types of memory, random access (RAM) and read-only (ROM).
- Data stored in RAM is lost when the power is turned off.

67
Computer Data

Data is organised and processed in different ways by different computers. Here we shall look at a simple 8-bit computer (like a Spectrum, for example).

Computer data

In chapter 22 we saw how numbers can be represented in **binary** using digital signals. A single binary unit is called a **bit** and will be either a logic 1 or a logic 0.

Of course computers must also be able to work with letters. This is done by using a **code**. The most common code is called the **ASCII code** and it uses groups of eight bits. Each letter of the alphabet, each numeral and even punctuation marks are all given their own combination of eight 1s and 0s. (See Figure 67.1.)

Each group of eight bits is called a **byte**.

character	ASCII code	value
A	0 1 0 0 0 0 0 1	65
r	0 1 0 0 1 0 0 0	72
?	0 0 1 1 1 1 1 1	63

Figure 67.1 Examples of the ASCII code

How data is stored in the memory

The memory can be thought of as a collection of 'pigeon holes', or **locations**. Each location contains eight bistables and so can store one byte (eight bits) of data. A 64K memory has over 64 000 different locations and can store a byte in each one.

Actually '1K' in computer terms means 2^{10} (or 1024), not 1000.

Address

Each memory location and each input and output device is given an **address** so that the CPU can find it. Codes are also used for addresses as well as for data. A typical **address code** is sixteen bits long (Figure 67.2).

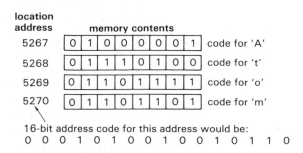

location
address   memory contents

5267   0 1 0 0 0 0 0 1   code for 'A'

5268   0 1 1 1 0 1 0 0   code for 't'

5269   0 1 1 0 1 1 1 1   code for 'o'

5270   0 1 1 0 1 1 0 1   code for 'm'

16-bit address code for this address would be:
0 0 0 1 0 1 0 0 1 0 0 1 0 1 1 0

Figure 67.2 The word 'Atom' could be stored in four memory locations

Sending data around the computer

Each part of the computer is connected by several sets of wires called **buses**. These carry the data, instructions about where the data is to go (e.g. which memory address) and what must be done with the data (e.g. write it into the memory). A typical arrangement is shown in Figure 67.3.

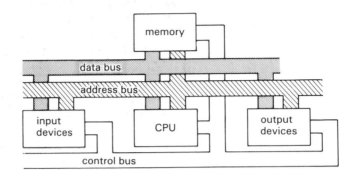

Figure 67.3   The bus network in a computer system

Figure 67.4   The data bus

Data bus *(Figure 67.4)*

The computer takes its data **one byte at a time** and keeps all the **bits** separate from each other. This means that the data bus must have eight separate wires.

Figure 67.5   The address bus

Address bus *(Figure 67.5)*

Since an address code is sixteen bits wide, the address bus which carries this code must have sixteen separate wires.

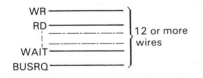

Figure 67.6   The control bus

Control bus *(Figure 67.6)*

This may contain twelve or more wires. Each one is used for a specific signal, e.g. WR means "write data into the memory".

How data is sent from the memory to the CPU

The CPU puts a 16-bit code on the address bus and puts a signal on the 'read' wire in the control bus.

The memory looks at the address code and works out which memory location is being requested. It then puts the data held at that address on the data bus. The CPU reads this data and then switches off the signal on the 'read' wire.

To **write** data into the memory a similar exchange of address codes is used but this time the CPU puts the data on the data bus and sends a control signal along the 'write' wire. The memory takes this data and puts it into the requested memory location.

171

Summary

- Information (letters, numbers, punctuation, etc.) is converted to data using an 8-bit code; 8 bits = 1 byte.
- Data is moved, one byte at a time, along the data bus.
- Memory locations are given an address.
- The CPU and memory are linked by an address bus as well as a data bus.
- A control bus carries control signals round the system.

68
Exercises

1 Although computers were originally used for **calculating**, they are now used for **storing and processing information, controlling other devices** as well as for **entertainment**.
   Choose *one* of these applications and use it to illustrate the following statements about computers:
   (a) "computers need some kind of input device"
   (b) "computers need some kind of output device"
   (c) "computers can store a large amount of information in a small space"
   (d) "some things are very difficult or impossible without computers"
   (e) "even computers cannot solve some problems"
   (f) "computers create new problems and difficulties".

2 Explain the terms 'bit' and 'byte'.

3 Explain how numbers and letters can be represented by binary digits using the ASCII code.

4 Name one high-level language. What advantages are there in writing programs in a high-level language rather than in simple machine code?

5 Data is held in the computer's memory.
   (a) What is the difference between the 'memory' and the 'store'?
   (b) What do the terms ROM and RAM mean?
   (c) Describe the differences between ROM and RAM in terms of: (i) the kinds of data held in each one (ii) how data is put into each one (iii) what happens when the power is turned off.

6 The CPU is linked to the rest of the system by three sets of connecting wires (or buses) (Figure 68.1):

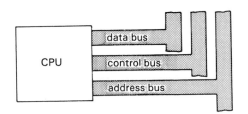

Figure 68.1

   (a) What signals are sent along a data bus?
   (b) What signals are sent along an address bus?
   (c) Name two uses for the control bus.

Section 9

Components

69
Introduction to Components

We have looked at electronic systems mainly as a combination of simple circuits used as building blocks. These circuits are made from **components**.

This section looks at the common components and how they behave. Use this section as a reference to look up details about a component if you meet it in a circuit and are not too sure what it is doing there.

Practical points
Electronics is best understood by actually building circuits and seeing how they behave. All the circuits shown in this book have component values marked on them and they will all work.

Component values
Components like resistors and capacitors have their **value** marked on them.

Transistors, integrated circuits and diodes do not have a value but they do have a **type number** marked on them. The best way of finding out what the different type numbers mean is to look at the manufacturers' **data sheets** or look in a good catalogue.

Component ratings
Apart from having a value, components are also **rated** by their manufacturers. The rating tells us the maximum current (and sometimes voltage) that the component can stand before the component overheats or the insulation breaks down.

Some components (like switches) are also given an **expected life-time**.

How many of these components can you identify?

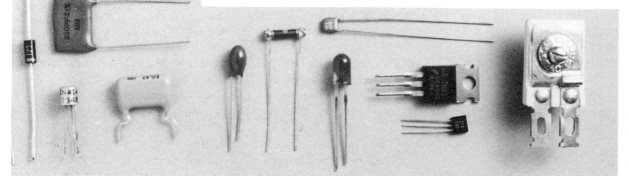

174

Colour codes

Rather than printing the value and tolerance on small resistors and capacitors, coloured bands are used as a code. The code is:

value bands	tolerance band	working voltage (capacitors only)
black = 0 brown = 1 red = 2 orange = 3 yellow = 4 green = 5 blue = 6 violet = 7 grey = 8 white = 9	no colour = ±20% silver = ±10% gold = ± 5%	red = 250 V yellow = 450 V blue = 630 V

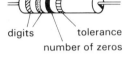

digits — tolerance — number of zeros

Figure 69.1 Reading the colour code on a resistor

How to read the colour code (resistors)

Hold the resistor with the bands nearest the left-hand side (Figure 69.1). Read the values from the left. The **first two bands** indicate the **first two digits**, the **third band** indicates the **number of zeros** and the **fourth band** indicates the **tolerance**. For example:

yellow, violet, red = 4700 (4.7 kΩ)
yellow, violet, orange = 47 000 (47 kΩ)
yellow, violet, brown = 470
yellow, violet, black = 47
brown, black, red = 1000 (1 kΩ)
brown, black, brown = 100
brown, black, black = 10

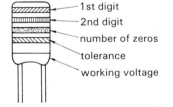

1st digit
2nd digit
number of zeros
tolerance
working voltage

Figure 69.2 Reading the colour code on a capacitor

Colour code for capacitors

Small capacitors (especially **polyester type**) use the same colour code. The first four bands are the same as for resistors. The last band indicates its **working voltage**. The values are in **picofarads**. (See Figure 69.2.)

Questions

1 Why is it important to know the rating of a component as well as its value?
2 Where would you look to find out more information about a particular component?
3 What is the colour code on the following resistors?
   (a) 270 kΩ ±5%   (b) 56 Ω ±10%
   (c) 10 MΩ ±20%   (d) 6.8 kΩ ±5%
4 What is the value of the resistors with these colour codes?
   (a) red, red, red, gold   (b) orange, white, brown
   (c) brown, black, red   (d) yellow, violet, green, silver

70
Resistors

These are components which are made with a known **resistance**. They are usually shaped like a tube. See Figure 70.1.

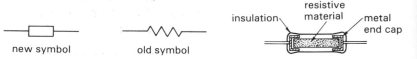

new symbol   old symbol

Figure 70.1   Resistors: symbols, and typical cross-section through a resistor

Markings

Sometimes the value is printed on the case but usually the colour code is used. (See chapter 69.)

Types

Several different **resistive materials** can be used, producing a variety of types of resistor. They all have their uses:

- **Carbon resistors** (Figure 70.2). Cheap, unstable (resistance changes with temperature and age), produce unwanted noise in circuits, power ratings up to 2 W. They can work with very high frequency signals and are used in many radio circuits.
- **Metal oxide resistors** (Figure 70.3). More expensive, accurate and stable than carbon types. They also produce less noise. Power ratings up to 0.5 W. These are used in precision equipment.
- **Wirebound resistors** (Figure 70.4). More expensive, can be very stable and accurate. They can have power ratings up to 25 W, but have a large case. Wirewound resistors are used in high precision equipment or high power circuits where large currents are flowing.

Figure 70.2   Typical carbon resistor

Figure 70.3   Typical metal oxide resistor

Figure 70.4   Typical wirewound resistor

Power rating

As explained in chapter 5, the heat produced in a resistor depends on the power used by the resistor. The **power rating** of a resistor describes the maximum power that it can handle before over-heating. Typical power ratings are 0.125 W, 0.5 W, 1 W, etc.

Safe design

For a safe design make sure that the power rating of each resistor in the circuit is large enough. Calculate this by power $= V^2/R$ (where V is the largest p.d. likely to be used), e.g. if a 100 Ω resistor is likely to have 6 V across it:

$$\text{power} = 6^2/100$$
$$= 36/100$$
$$= 0.36 \text{ W}$$

A resistor rated at 0.25 W would be too small, so choose a 0.5 W resistor for safety.

See chapter 80 for more details on how to calculate power.

Preferred values

Manufacturers sell only a limited range of resistor values. These are called **preferred values**. There are 12 resistor values between $1\,\Omega$ and $10\,\Omega$; 12 more between $10\,\Omega$ and $100\,\Omega$, etc. This is called the **E12 series**.

1.0	1.2	1.5	1.8	2.2	2.7	3.3	3.9	4.7	5.6	6.8	$8.2\,\Omega$
10	12	15	18	22	27	33	39	47	56	68	$82\,\Omega$
100											$820\,\Omega$
$1.0\,k\Omega$											$8.2\,k\Omega$
$10\,k\Omega$											$82\,k\Omega$
$100\,k\Omega$											$820\,k\Omega$
$1.0\,M\Omega$											$8.2\,M\Omega$
$10\,M\Omega$											$82\,M\Omega$
etc.											

In most circuits the exact value of the resistor is not too important. If, for example, a $440\,\Omega$ resistor is needed then a $470\,\Omega$ would probably work just as well. If an exact $440\,\Omega$ is needed then two $220\,\Omega$ could be used in series.

Tolerance (or accuracy)

A resistor is unlikely to be exactly the value marked on it. The **tolerance** of a resistor tells us how close its actual value was to its marked value when it was made. For example, a $470\,\Omega$ resistor with a tolerance of $\pm10\%$ could have a value of between $423\,\Omega$ (i.e. $470\,\Omega - 47\,\Omega$) and $517\,\Omega$ (i.e. $470\,\Omega + 47\,\Omega$).

Typical tolerances are $\pm5\%$, $\pm10\%$ and $\pm20\%$, and each individual resistor sold is marked accordingly.

Writing resistor values

A system of numbers and letters is often used to indicate value and tolerance. The code is a British Standard (1852). This is best described by example:

values	written as:	tolerance
$47\,\Omega$	47R	$F = \pm1\%$
$470\,\Omega$	470R (or K47)	$G = \pm2\%$
$4.7\,k\Omega$	4K7	$J = \pm5\%$
$47\,k\Omega$	47K	$K = \pm10\%$
$47\,k\Omega$	470K (or M47)	$M = \pm20\%$
$4.7\,M\Omega$	4M7	
$47\,M\Omega$	47M	

So, **6K8J = 6.8 kΩ $\pm$5%**.

Summary of uses of resistors
● To limit current flowing into devices (Figure 70.5).
● To provide a potential difference (Figure 70.6).
● Two resistors in series make a **potential divider** (Figure 70.7).

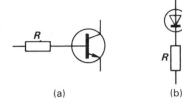

(a)   (b)

Figure 70.5 (a) Resistor limits the current flowing into the base of the transistor   (b) Resistor keeps the current flowing into the LED small

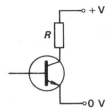

Figure 70.6 A change in the collector current produces a change in the p.d. across R

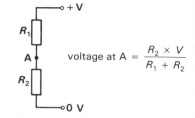

voltage at A $= \dfrac{R_2 \times V}{R_1 + R_2}$

Figure 70.7 The voltage at A depends on the resistor values

Questions
1   What type of resistor has the largest power rating?
2   Which of the following resistors are *not* in the E12 series:
    (a) 270 kΩ   (b) 39 Ω   (c) 42 kΩ   (d) 10 MΩ   (e) 6.9 kΩ
3   Which is the nearest resistor in the E12 series to these?
    (a) 54 kΩ   (b) 4.2 kΩ   (c) 96 MΩ   (d) 35 Ω   (e) 730 kΩ
4   In each of the circuits in Figure 70.8, is the resistor being used to limit the current or to provide a p.d?

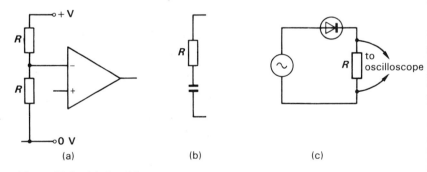

(a)   (b)   (c)

Figure 70.8 (a) Providing the reference voltage for a comparator   (b) Part of a monostable circuit   (c) Circuit to demonstrate the action of a diode

5   Is the voltage at point A 'high', 'low' or 'about half the supply voltage' in each of the potential dividers in Figure 70.9?

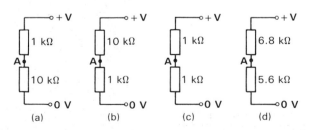

(a)   (b)   (c)   (d)

Figure 70.9

71
Variable Resistors

Potentiometers or **variable resistors** use a **track** of a fixed resistance and a **slider** which can move over the track. See Figure 71.1.

Maximum resistance between A and S is when the slider is at the far right of the track (point B).

Markings

The value (i.e. resistance of the track) is marked on the case together with the letters 'LIN' or 'LOG'. This refers to how the resistance changes round the track. See Figure 71.2.

Figure 71.1 Potentiometer (a) Symbol (b) Construction of a typical device

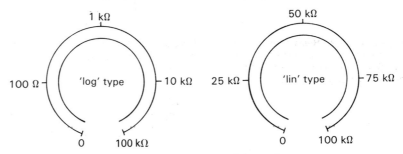

Figure 71.2 How the resistance changes round the track

Power rating

Like any other resistor, the **power rating** indicates the maximum power which can be handled safely. A typical rating is 3 W.

Types of variable resistor or potentiometer

Open (or skeleton) preset

This is useful for the initial setting up of a circuit.

Standard rotary

This is used when continual adjustments need to be made (e.g. volume controls).

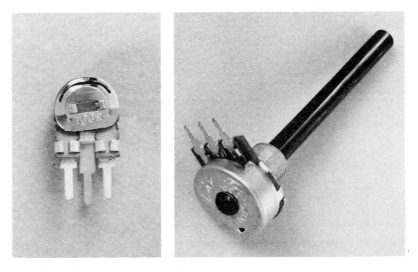

Typical examples of an open preset variable resistor (*left*) and a standard rotary resistor (*right*)

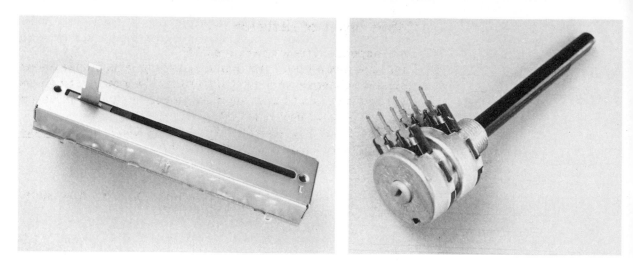

Typical example of a slider-type variable resistor (*left*) and a dual gang rotary resistor (*right*)

Slider

This is a 'straightened out' version. It is easier to tell at a glance what the setting is.

Dual gang (or tandem)

This is simply two separate variable resistors sharing the same spindle. They are used when two circuits need to be adjusted at the same time, e.g. in stereo audio systems.

Uses of potentiometers

- If we use just two connections (the slider and one of the track connections) we have a variable resistor which can be used in place of a normal resistor (Figure 71.3). This is most useful in the initial setting up of a circuit when we are not sure of the exact value of the resistor needed.

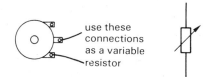

use these connections as a variable resistor

Figure 71.3

- If we use all three connections, the potentiometer can be used as a variable potential divider (or **pot**) (Figure 71.4). The voltage on the slider drops as it is moved downwards. An example of this use is a **volume control**.

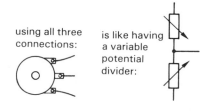

using all three connections:

is like having a variable potential divider:

Figure 71.4

Figure 71.5 Light-dependent resistor (LDR): symbol and typical example

Other types of resistor

Light-dependent resistor (*Figure 71.5*)

These are made from a material (e.g. cadmium sulphide) whose resistance changes when light shines on it. In complete darkness the resistance may be as high as 10 MΩ. In full sunlight the resistance may be as low as 1 kΩ.

A light-dependent resistor (LDR) can be used in a potential divider to make a **light sensor**. See chapter 7.

Thermistors

These are resistors whose values depend on the temperature. There are two types (Figure 71.6):

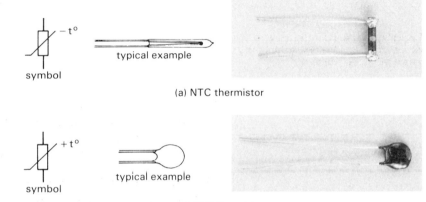

(a) NTC thermistor

(b) PTC thermistor

Figure 71.6   Thermistors: symbols and typical examples

type	resistance	
	at 20°C	at 100°C
RA53 (NTC)	5 kΩ	100 Ω
VA1026 (NTC)	400 Ω	30 Ω
PTC	100 Ω	1 kΩ

Figure 71.7   Behaviour of typical thermistors

PTC (positive temperature coefficient) thermistors have a **high** resistance when hot. They are used to protect amplifiers and other circuits by reducing the current flowing if the circuits overheat.

NTC (negative temperature coefficient) thermistors have a **low** resistance when hot. They can be used in a potential divider to make a **heat sensor**.

See Figure 71.7 for typical values of thermistors.

72

Capacitors

Construction

A capacitor is made from two sheets of metal foil (called **plates**) separated by a sheet of insulator (or **dielectric**). Usually they are rolled into a tube. See Figure 72.1.

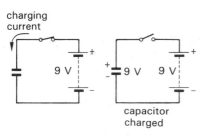

Figure 72.1 Capacitors (a) Symbols (b) Construction of a typical device

Figure 72.2 A capacitor can store charge

When a capacitor is connected across a p.d., current flows into the plates for a short time. The capacitor becomes **charged up**. It will store this charge, even if the source of the p.d. is removed. The capacitor will still have a p.d. across it because of the charge now stored on its plates. See Figure 72.2.

Values of capacitors

The amount of charge stored by a capacitor depends on the p.d. and the **capacitance** of the capacitor. Capacitance is measured in **farads**. A farad is very large. More common units are microfarads (μF), nanofarads (nF), and picofarads (pF).

$$1000\,pF = 1\,nF$$
$$1000\,nF = 1\,\mu F$$

Capacitors and direct current

Although no current can actually flow through the capacitor, some current will flow through the circuit while the capacitor charges or discharges.

If the capacitor is connected in series with a resistor, less current can flow. Charging and discharging take longer if the capacitor and resistor are both large (Figure 72.3).

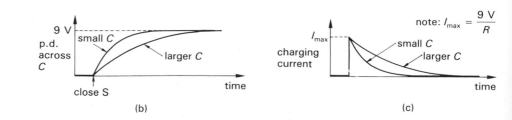

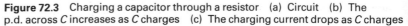

Figure 72.3 Charging a capacitor through a resistor (a) Circuit (b) The p.d. across C increases as C charges (c) The charging current drops as C charges

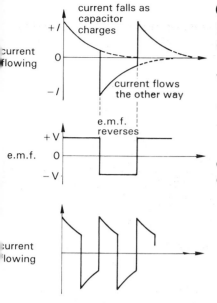

Figure 72.4 (*top*) Current flow in an RC circuit when a square wave e.m.f. is applied. (*bottom*) The current will stay high if the frequency of the a.c. is high.

Capacitors and alternating current

If a.c. is used the capacitor may never get fully charged or discharged before the current changes direction. (Figure 72.4)

Some current will always be flowing in the circuit. The current will stay **high** if the **frequency** of the a.c. is high.

Types of 'fixed value' capacitor and their markings

Normal capacitor (*Figure 72.5*)

● The **silver mica** types are very accurate and stable.
● The **polyester** types are 'general purpose' components.

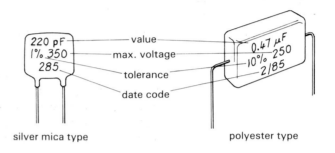

silver mica type   polyester type

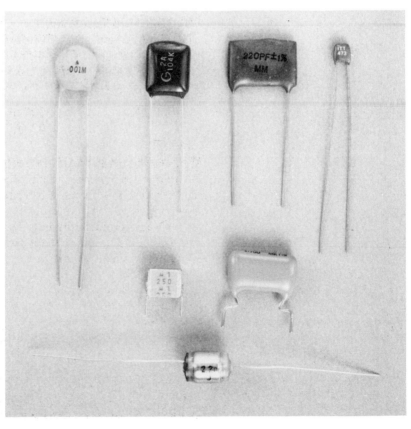

Figure 72.5 Typical normal capacitors (connect either way round)

Electrolytic capacitor (*Figure 72.6*)

● These are specially made to have a large value in a fairly small case.
● They are not very good for use with high frequency signals.
● They must be connected the right way round. The case is marked to show this.
● **Tantalum bead** capacitors behave like electrolytics. They are more accurate, have a very small case and a low working voltage.

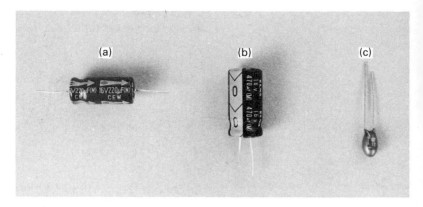

Figure 72.6 Polarised capacitors (must be connected the right way round) (a) Axial lead electrolytic (b) Radial lead electrolytic (c) Tantalum bead

Variable capacitors

The value of a capacitor depends on the area of the plates and the distance between them, as well as the type of dielectric used. Figure 72.7 shows two sorts of variable capacitor.

Figure 72.7 Variable capacitors (a) Air spaced. Used in tuner of radios. (b) Trimmer. Used in initial setting up of circuit.

● **Air spaced**. The capacitor is varied by moving one set of plates. This alters the overlap area of the plates.
● **Trimmer**. The capacitor is varied by squeezing the plates closer together.

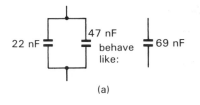

22 nF ⊣⊢ 47 nF behave like: ⊣⊢ 69 nF

(a)

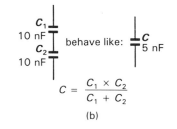

C_1 10 nF behave like: C 5 nF
C_2 10 nF

$$C = \frac{C_1 \times C_2}{C_1 + C_2}$$

(b)

Figure 72.8 (a) Two capacitors in parallel add together (b) Two capacitors in series give less capacitance

Capacitors in series and in parallel *(Figure 72.8)*

Two capacitors in parallel add together to give a larger capacitance. Two capacitors in series give less capacitance. This is the opposite to combinations of resistors.

Summary of the uses of capacitors

● To block a d.c., but allow an a.c. signal to pass (Figure 72.9).
● To let high frequency signals pass easier than low frequencies (Figure 72.10).

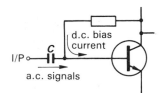

Figure 72.9 A blocking or coupling capacitor is used on the input to an amplifier

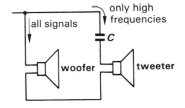

Figure 72.10 A simple crossover uses a capacitor to allow only high frequency signals through to the tweeter

● To give a time delay when being charged through a resistor (Figure 72.11).
● To store a charge and therefore smooth a changing d.c. (Figure 72.12).

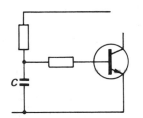

Figure 72.11 A simple monostable uses the time delay while a capacitor charges up

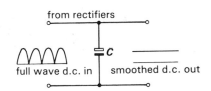

Figure 72.12 A power supply uses a smoothing capacitor

● With an inductor to make a tuned circuit (Figure 72.13).

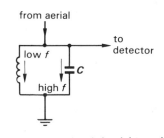

Figure 72.13 A tuned circuit is used in a radio receiver

Questions

1 What units are the values of capacitors measured in? Which of these units is the largest?
2 Why does current never flow 'through' a capacitor?
3 Look at the circuit in Figure 72.3.
   (a) Which of the following affects the time the capacitor takes to charge up when S is closed? (i) the value of the resistor (ii) the value of the capacitor (iii) the p.d. used.
   (b) Draw a graph showing how the voltage at point A changes when the switch is closed.
   (c) Name one type of circuit which uses this time delay.

73
Inductors

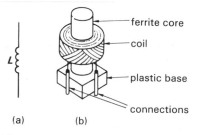

Figure 73.1 Inductors (abbreviation *L*) (a) Symbol (b) Typical example of a small (r.f.) inductor

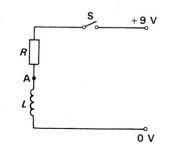

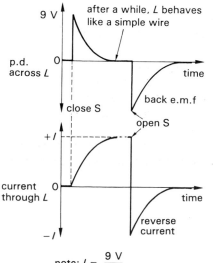

note: $I = \dfrac{9\ V}{R}$

Figure 73.2 Behaviour of an inductor to a sudden change in e.m.f.

Construction

The simplest inductor is a coil of wire. Often this coil is wound around a **core** of either **iron** or **ferrite**. (Ferrite is a brittle material containing iron oxide.) See Figure 73.1.

How an inductor behaves

We might expect a current to flow in a coil the instant that a p.d. is put across it. This is not the case, however (see Figure 73.2).

Because of an effect called **self inductance**, the current takes some time to build up to the full value after switching on the supply.

When we switch off the supply the coil produces its own e.m.f. for a short while. This **back e.m.f.** is in the other direction. A reverse current flows for a while.

The time taken is longest when the value of the resistor and the **inductance** of the coil are large.

Inductors and d.c.

The current does not take very long to build up and so for d.c. circuits the coil soon behaves like a simple resistance.

Back e.m.f.

Components which contain a coil (e.g. a relay) will produce a sudden **reverse p.d.** or **back e.m.f.** when they are switched off. It can damage other components in the circuit.

Inductors and a.c.

If we try to send an alternating current through an inductor, then the inductance effect will mean that the current takes time to build up. If the frequency of the a.c. is high then the current never has time to build up to the maximum before it has to change direction again. (Figure 73.3) **Inductors cannot pass high frequency a.c. signals.**

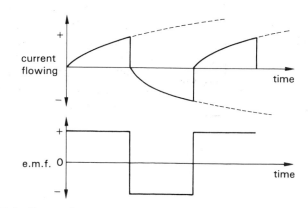

Figure 73.3 Current flow through an inductor when a square wave signal is applied to it

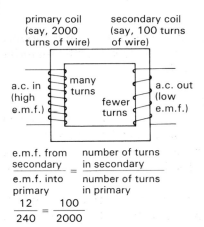

primary coil
(say, 2000
turns of wire)

secondary coil
(say, 100 turns
of wire)

a.c. in
(high
e.m.f.)

many
turns

fewer
turns

a.c. out
(low
e.m.f.)

$$\frac{\text{e.m.f. from secondary}}{\text{e.m.f. into primary}} = \frac{\text{number of turns in secondary}}{\text{number of turns in primary}}$$

$$\frac{12}{240} = \frac{100}{2000}$$

Figure 73.4 e.m.f. into and out from a transformer

Inductors and capacitors

We can see that inductors behave in the opposite way to capacitors. In fact inductors and capacitors are often used together in tuned circuits and filters.

Transformers

A transformer is used to change an e.m.f. without much loss of energy. It consists of at least two coils wound around an iron **core**.

How it behaves

When an alternating e.m.f. is put across the primary coil, an a.c. flows. This induces another a.c. in the secondary coil. The e.m.f. behind this secondary current depends on the number of turns in the two coils. (Figure 73.4)

Step-up and step-down transformers

A **step-up** transformer has **more** turns in the secondary than in the primary coil. The transformer produces a **larger** e.m.f. and is used to produce the high voltages needed inside TV tubes.

A **step-down** transformer has **less** turns in the secondary than in the primary coil. It produces a **smaller** e.m.f. and is used in low-voltage power supplies to provide a low voltage from the 240 V mains.

Ratings of transformers

The VA rating of a transformer describes how much current it can deliver at its output e.m.f. before overheating. For example a transformer which can deliver 2 amps at 5 V is rated at 10 VA.

Questions

1. What is the symbol for an inductor?
2. What happens when we suddenly disconnect a p.d. from the ends of an inductor?
3. Why must we be careful when using relays in circuits?
4. Does an inductor allow high frequency a.c. signals through?
5. Look at the circuit in Figure 73.5 and describe what happens if we steadily increase the frequency of the signal going in.
6. Describe two uses of an inductor.
7. What is a step-down transformer and where might it be used?
8. How much current can a 5 VA transformer deliver if its output e.m.f. is 20 V?

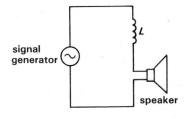

signal generator

L

speaker

Figure 73.5

74
Transistors

Construction and types

Bipolar transistors are made from two different types of silicon (**p-type** and **n-type**). The exact method used varies but a typical example is shown in Figure 74.1. They can be **n-p-n** or **p-n-p** type, depending on how the silicon is arranged. (See Figure 74.2.) Nearly all the circuits in this book use n-p-n transistors. **Field-effect transistors** (FET) are made differently.

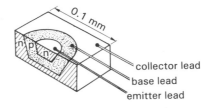

Figure 74.1 Construction of a typical n-p-n bipolar transistor. Although the actual transistor is about 0.1 mm across it is mounted inside a larger plastic case.

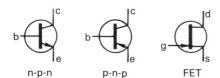

Figure 74.2 Transistor symbols

How bipolar transistors behave

When connected in a circuit, as shown in Figure 74.3, current can only flow into the collector if a small current flows into the base as well. Both currents flow out through the emitter.

When no current flows into the base, the collector is at +9 V.

Because of the way it is made, the voltage at the base must be about 0.6 V before I_C can flow. Then V_C starts to drop.

If the base is at more than about 0.7 V, then I_C is so large that the voltage at C is zero.

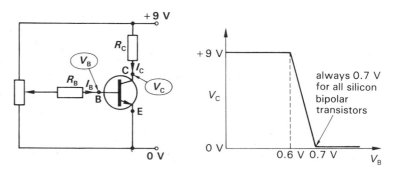

Figure 74.3 Investigating the behaviour of a bipolar transistor

Current gain

The current gain of the transistor is a comparison between the size of the collector current (I_C) and the base current (I_B):

$$\text{current gain } (h_{FE}) = I_C/I_B$$

For a typical bipolar transistor (say a BC108), $h_{FE} = 100$.

How field-effect transistors behave *(Figure 74.4)*

The FET behaves like a voltage-controlled variable resistor. The current flowing through the drain down to the source (I_{DS}) depends on the voltage we put on the gate. We can reduce I_{DS} by making the p.d. between G and S negative.

Unlike bipolar transistors, almost no current flows into the gate. It has a very high input resistance.

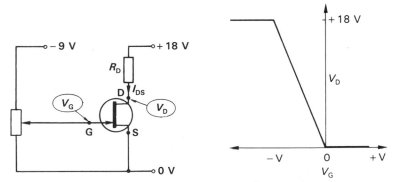

Figure 74.4 Investigating the behaviour of a field-effect transistor

Using transistors

● A bipolar transistor can be used as a switch (Figure 74.5). If the base voltage (V_B) is low, no base current flows and the transistor is turned off. The collector voltage (V_c) is high (9 V). If V_B is high (more than 0.7 V) the transistor is turned on. Then V_c is low. This can be used to control a load.

● It can also be used as an amplifier (Figure 74.6). The transistor is turned half-way on (or biassed) by current flowing through R_B, and the voltage at the collector (V_c) is at half the supply voltage. A signal applied to the base will then either turn it further on or off. This causes a large change in V_C.

● A field-effect transistor can be used as an amplifier (Figure 74.7). The gate is biassed to be at a negative voltage. A changing voltage signal applied to the gate changes the current flowing through R_D and so produces an amplified voltage signal at D.

The FET amplifier has a very high input resistance and is ideal for signal sources which cannot deliver very much current.

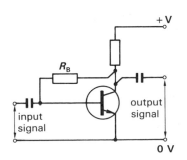

could be a bulb, for example

sensor

may respond to changes in light, for example

Figure 74.5 Using an n-p-n transistor as a switch

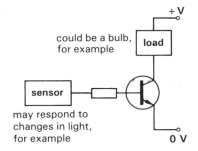

Figure 74.6 Using an n-p-n transistor as an amplifier

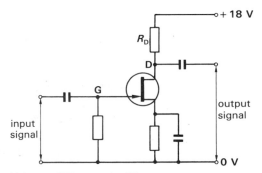

Figure 74.7 Using an FET as an amplifier

189

A heat shunt helps to protect transistors when soldering.

Practical points
● Connecting the transistor the wrong way round (say base to 0 V rail and emitter to a positive voltage) will instantly ruin it.
● The transistor can be damaged by too much heat, either by:
  1 careless use of the soldering iron, or
  2 allowing too much current to flow. Manufacturers always state the maximum current (I_c *max*) that can safely flow through it.

Questions
1 Explain what the 'current gain' of a transistor means.
2 Draw a circuit which will let you measure the voltages round a transistor. Explain the results that you would hope to get.
3 Explain how a transistor can be used as a switch.
4 How can a transistor be damaged?

75
Diodes

Construction

A diode is made from two types of silicon or germanium (p-type and n-type). The exact construction varies but a typical example is shown in Figure 75.1. The case of a diode is marked to show which end is the cathode. (See Figure 75.2.)

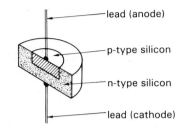

Figure 75.1 Section through a typical diode

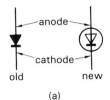

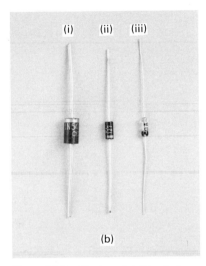

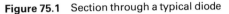

Figure 75.2 Diodes (a) Symbols
(b) Typical devices: (i) silicon rectifier
diode (ii) small signal diode (iii)
glass-cased germanium detector diode

How a diode behaves

If a diode is connected to a p.d., its behaviour depends on which way round it is and how much p.d. we apply. (See Figures 75.3 and 75.4.)

Important points

● The manufacturer will state the **maximum reverse voltage** (V_{RRM}) before diode will break down and conduct.
● The manufacturer will also state the **average forward current** ($I_F ave$) that the diode can safely conduct.
● There will always be a **forward voltage drop** of about 0.7 V across a silicon diode when it is conducting in forward bias.
● **Germanium** diodes only have a forward voltage drop of about 0.2 V but they cannot stand as much current as silicon diodes.

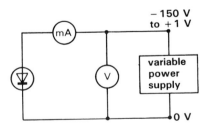

Figure 75.3 Arrangement to investigate the behaviour of a diode

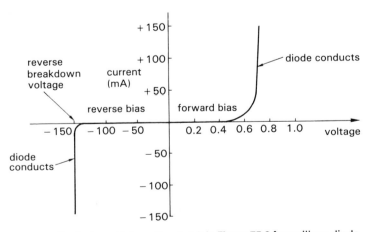

Figure 75.4 Typical results from the circuit in Figure 75.3 for a silicon diode. This graph is sometimes called the characteristic of the diode.

Figure 75.5 gives a summary of the behaviour of a silicon diode.

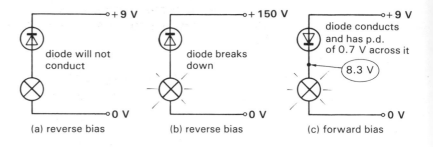

Figure 75.5 Summary of the behaviour of a silicon diode. Note that the diode is not damaged when it breaks down, it simply starts conducting.

(a) reverse bias (b) reverse bias (c) forward bias

Other types of diode

Zener diodes behave like normal diodes except that they break down at a much lower reverse voltage (say 5.6 V). They are used, in reverse bias so that they break down, in **voltage regulators** (Figure 75.6).

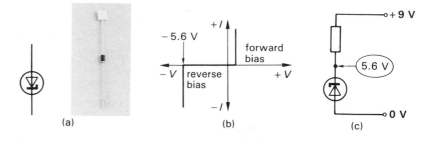

Figure 75.6 Zener diodes (a) Symbol and typical device (b) Characteristic for a 5.6 V Zener diode (c) Simple circuit to produce +5.6 V from a +9 V supply

(a) (b) (c)

Light-emitting diodes or LEDs have a transparent case (often red). When connected in forward bias they give out a small amount of light and are used in displays and as indicators. They have a forward voltage drop of about 2 V and a resistor must be connected in series to keep the current small (about 15 mA) (Figure 75.7).

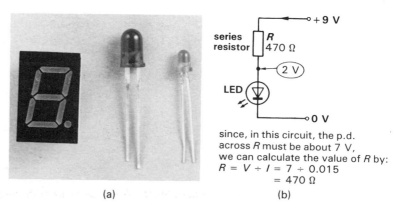

Figure 75.7 LEDS (a) Examples of LEDs (b) LED in circuit with its series resistor

since, in this circuit, the p.d. across R must be about 7 V, we can calculate the value of R by:
$R = V \div I = 7 \div 0.015$
$= 470\ \Omega$

(a) (b)

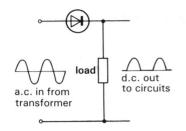

Figure 75.8 A single diode used as a half-wave rectifier

Summary of the uses of diodes

● To turn an a.c. into a d.c. in power supplies (Figure 75.8).
● To protect circuits against accidental damage if the power supply is connected the wrong way round (Figure 75.9).

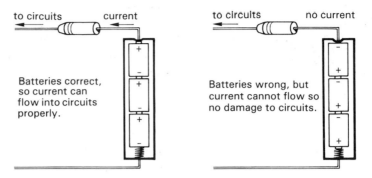

Figure 75.9 A protection diode is connected to the battery holder

● To protect transistors from damage caused by the back e.m.f. when a relay is turned off (Figure 75.10).

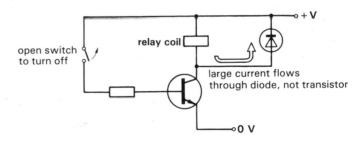

Figure 75.10 Without a protection diode the transistor would be damaged by the large current produced by the coil when it turns off

● In a **detector** in simple radio circuits to separate the a.f. signal from the r.f. carrier (Figure 75.11).

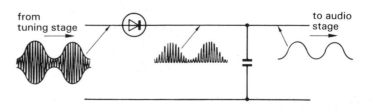

Figure 75.11 The diode removes the negative half of the wave

● In a **matrix** so that digital signals can be directed to the right point in a circuit (Figure 75.12).

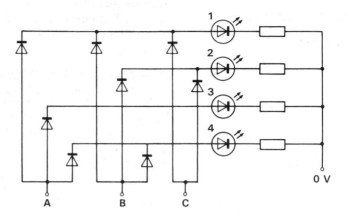

Figure 75.12   In this matrix, when A is given +5 V, then the LEDs 1, 3 and 4 light up

Questions

1   How much p.d. is needed for a diode to conduct in forward bias?
2   What happens when a reverse breakdown voltage is applied to a diode? What sort of diode is used in this way?
3   What precaution must we take when using LEDs?
4   Draw simple circuits to show how diodes can be used for protecting other components.
5   Look at the diode matrix in Figure 75.12.
    (a) Which LEDs will light up when point B is given +5 V?
    (b) Which LEDs will light up when point C is given +5 V?

76
Integrated Circuits

Integrated circuits (or **silicon chips**) are made by specially treating small areas of a piece of silicon to form the minute transistors, diodes, resistors and interconnections which together make up a circuit. The circuit is then put in a case.

Connection to the integrated circuit is made via pins coming out of the case.

Figure 76.1 A 16-pin DIL IC. The device shown here contains logic gates.

Figure 76.2 A QIL packaged IC. This device is an audio amplifier.

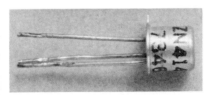

Figure 76.3 A metal-cased IC

A cutaway of a typical integrated circuit. Most of the space is taken up by the case and connecting leads.

Types of case

DIL (or **dual-in-line**) is the most common type. The pins are all in line down each edge of the case. Figure 76.1 shows a 16-pin DIL case. Other common sizes are 8, 14, 28 and 40 DIL.

QIL (or **quad-in-line**) are sometimes found (Figure 76.2). The pins are offset to give four rows. The large metal tabs shown are a simple heat sink.

Metal-cased ICs can be confused with transistors. The device shown in Figure 76.3 (ZN414) in fact contains most of the circuitry for an AM radio.

Figure 76.4 An IC socket

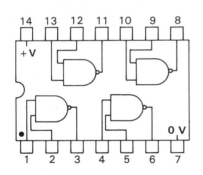

Figure 76.5 Pinout of a 7400

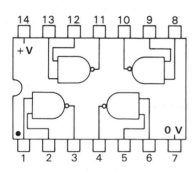

Figure 76.6 Pinout of a 4011

IC sockets

Shown in Figure 76.4 is an IC socket. It is soldered to the circuit board and then an IC is pushed into the holes on the top. These avoid damaging ICs when soldering and make it easy to change a faulty IC if needed.

Markings

The IC will be marked with its **type number** (e.g. 555). Other markings include the manufacturer's code (e.g. SN is used by Texas Industries) and a date code to indicate when it was made.

Types of digital ICs

There are two main types (or **families**) of digital ICs; **TTL** (which start with the numbers '74') and **CMOS** (which start with '40' or '45').

Most of the common packages of gates, latches and decoders are available in either type. For example, two packages which contain four NAND gates are shown in Figures 76.5 and 76.6.

The TTL type (Figure 76.5) can work quickly but takes more current than CMOS. It needs a +5V power supply.

The CMOS type (Figure 76.6) takes less current but cannot work as quickly as TTL. It works with any supply between +3V and +15V.

Practical points when using ICs

Some ICs are easily damaged by overloading and can be affected by sudden changes in the power supply. Take care with digital ICs:

● Connect all unused input pins to the 0V or positive rail.
● Connect a small capacitor across the power supply rails.
● Never connect an input signal when the power supply is off.
● Never connect a signal larger than the power supply voltage.

Current ratings

Most ICs (apart from the large power amplifiers and voltage regulators, etc.) cannot stand much current flowing from their outputs; a typical value might be 50mA. If they are to be used to control large devices they can be interfaced via a **transistor buffer** (chapter 26) which can stand more current.

Questions

1 What do the following abbreviations stand for?
   (a) IC   (b) DIL   (c) QIL
2 Why is it not a good idea to solder ICs directly in place?
3 What are the main differences between TTL and CMOS ICs?
4 Look in a components catalogue and find the type number of:
   (a) an A to D converter   (b) a CPU
   (c) a BCD-to-decimal decoder (d) a + 12V voltage regulator
   (e) an analogue switch   (f) a stereo preamplifier

77
Meters

Three things are commonly measured in electronic circuits:
- potential difference (using a voltmeter)
- current (using an ammeter)
- resistance (using an ohmmeter).

Figure 77.1 A typical multimeter.

Multimeters

A **multimeter** is a very useful piece of test equipment. It can be used as an ammeter, voltmeter or ohmmeter. Switches are used to change between these functions. The trade name of the make of multimeter shown in Figure 77.1 is 'AVO'.

Measuring voltage *(Figure 77.2)*

The multimeter is set to measure volts and is connected across the p.d. we want to measure. If the bottom lead is on the 0 V rail, the meter will show the **voltage at the top lead**.

Measuring current *(Figure 77.3)*

The multimeter is set to read current and is connected so that the current has to flow through it. Usually this means cutting a wire and putting the ammeter in the gap.

Measuring resistance *(Figure 77.4)*

An ohmmeter measures resistance by measuring the current flowing from its own, internal, battery. Notice that the scale is 'backwards', i.e. a **low resistance** lets a **large current** flow.

Because the e.m.f. of the internal battery can change with age, an ohmmeter is always **zeroed** before use. To do this we connect the leads together (i.e. zero resistance) and adjust a variable resistor until the meter reads exactly zero.

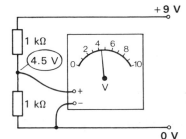

Figure 77.2 Measuring the voltage at the centre of a potential divider

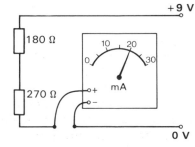

Figure 77.3 Measuring the current flowing through a circuit

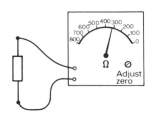

Figure 77.4 Measuring the resistance of a component

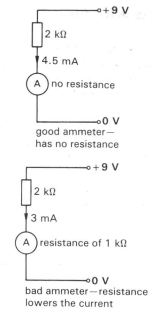

2 kΩ

4.5 mA

(A) no resistance

good ammeter —
has no resistance

2 kΩ

3 mA

(A) resistance of 1 kΩ

bad ammeter — resistance
lowers the current

Figure 77.5 The effect of meter resistance on current measurements

A digital multimeter. The large display makes it very easy to use. What other advantages does it have?

Impedance of meters

The meter will have some impedance (or resistance). The impedance of ammeters and voltmeters is very different.

Ammeter *(Figure 77.5)*

A good ammeter has a very **small** impedance so that it does not reduce the current it is measuring.

Voltmeter *(Figure 77.6)*

A good voltmeter has a very **high** impedance. This means that it will take very little current from the circuit. If the impedance is small, then we get a reading which is lower than it should be. This is why a low impedance voltmeter cannot be used to measure the p.d. across a high-value resistor.

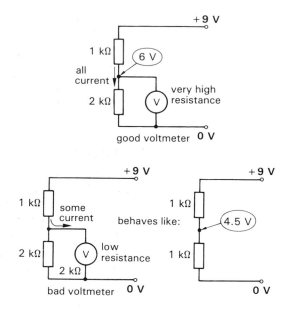

Figure 77.6 The effect of meter resistance on voltage measurements

Analogue and digital meters

An **analogue meter** (like the one in Figure 77.1) uses a fixed scale and a moving pointer to indicate the reading. They are useful when trying to spot changes in current or voltage (e.g. a slow a.c.). Most analogue voltmeters do not have an extremely high impedance and should be used with this in mind.

A **digital meter** (see chapter 27) gives a reading on 7-segment displays. They take some time to make their measurements and so are not very useful for changing currents or voltages. Even a fairly cheap digital voltmeter has a very high impedance.

Testing common components with an ohmmeter

The best type of meter to use for testing components is the analogue meter. The component must be taken out of the circuit.

The test usually involves measuring the resistance twice, once with the leads one way round then again with the leads reversed:

component	resistance with leads normal	resistance with leads reversed
fuses:		
good	0	0
blown	infinite	infinite
diodes:		
good	low	high
shorted out	low	low
open circuit	high	high
capacitors:		
good electrolytic	starts low then goes high either way round	
leaky electrolytic	starts low and stays low either way round	
good 'normal' type	infinite	infinite
good transistors:		
base to emitter	high	low
base to collector	high	low
emitter to collector	high	high

Questions

1   Suppose we want to measure the p.d. across a resistor and the current flowing through it. Draw a circuit diagram showing how to connect a voltmeter and an ammeter.
2   Should a good voltmeter have a low or a high resistance?
3   Which type of meter is best for making accurate readings of a steady p.d. across a high-value resistor?
4   What do you think 'AVO' stands for?
5   What must you do to an ohmmeter before trying to measure a resistance?

78
Plugs, Sockets and Cables

A Euro connector. The socket is on the mains lead. The plug is on the equipment.

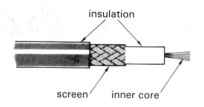
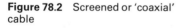

Figure 78.2 Screened or 'coaxial' cable

Plugs and sockets allow the building blocks in systems to be connected together easily and allow us to remove part of the system when it needs servicing or adjusting. To be reliable the plug, socket and cable must be able to stand the amount of current flowing through them and not be a source of **noise**, either by moving when in use or by allowing interference to enter the system.

Mains connectors

The familiar **13 amp plug** is shown in Figure 78.1. Note that:

● The fuse must be the right value.
● The cable grip must be tight enough to stop the wires from being pulled loose.

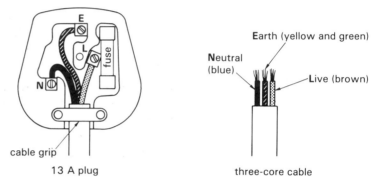

Figure 78.1 13 A mains plug

Mains cables generally have three **cores** (wires); live, neutral and earth. Each core is made from many thin wires so that there is less chance of it breaking.

Many items of equipment use the **Euro connector**. These plugs are moulded on to the cable.

Audio system connectors

One of the most common sources of noise in a system is caused by it picking up unwanted waves from radio transmissions and also from the 240 V a.c. mains. This can be reduced by putting an earthed, metal **screen** around the whole system. The circuits can go inside a metal case, but what about the connecting leads?

Screened cable

The answer is to make one of the wires in the connecting lead into a **braid** or **mesh**, surrounding (but insulated from) the other cores. This braid is then connected to the earth at one end, so making a kind of flexible **screen** round the other cores. It is also called **coaxial cable**. (Figure 78.2)

Very often one of the connecting wires will be the 0 V rail so the braid screen can do two jobs at once. (See Figure 78.3.)

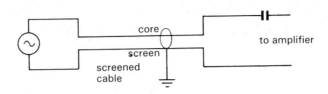

Figure 78.3 Using screened cable to connect a signal source to an amplifier

Figure 78.4 Screened plugs and sockets (a) BNC connectors (b) Coaxial connectors (c) Phono and DIN (d) UHF connectors.

Screened plugs and sockets

There are various types of plugs and sockets for use with screened cables. They are all designed to keep the screen around the connection right up to where it enters the case.

Screened cables should always be used where the signal is small and about to be amplified. Some examples are shown in Figure 78.4.

● **Phono** and **DIN connectors** are used between the amplifier and an input device (e.g. record pick-up) of an audio system.
● **Co-ax connectors** are used to connect a TV aerial lead.
● **BNC connectors** are used at the input to an oscilloscope.
● **UHF connectors** are used for UHF signals, e.g. in videos. The screw thread holds the plug and socket firmly together.

(a)

(b)

(c)

(d)

Where not to use screened cables

Screened co-ax cables have **capacitance**. The effect is like putting a capacitor across the wires; it will allow a.c. signals across. They should never be used to carry large alternating currents. For example:

- Screened cables must not be used as mains leads.
- Screened cables must not be used to connect an amplifier to a speaker system. Thick **two-core** cable is best.

Questions

1   What is the correct type of cable, plug and socket to:
    (a) connect a tape recorder to an amplifier
    (b) connect a power supply to the mains
    (c) connect an amplifier to its speakers?
2   What two jobs is the braid screen in this cable doing? (Figure 78.5)

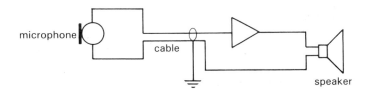

Figure 78.5

3   Why is each core of a mains cable made from many thin wires rather than from one thick wire?

79
Switches and Relays

A **mechanical switch** contains metal **contacts** which can be **opened**, preventing current from flowing, or **closed**, allowing current to flow. The **current rating** of a switch indicates the maximum current that can be switched before the contacts are damaged. The simplest type (a **one-way switch**) is shown in Figure 79.1.

Two-way switch (or single pole/double throw: SPDT)
This allows **one input** to be connected to either of **two outputs**. It would be used, for example, to select either a bell or a light in an alarm system. (See Figure 79.2.)

Figure 79.1   An on/off toggle switch

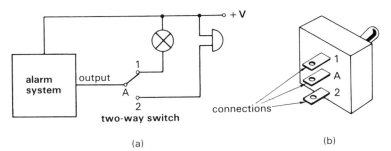

(a)   (b)

Figure 79.2   (a) Using a two-way switch to select the output device for an alarm system   (b) Rear view of a typical two-way switch

Two-pole switch (or double pole/single throw: DPST)
This is really two separate switches controlled by one lever. A typical use is as a mains switch, which disconnects both live and neutral at the same time. (Figure 79.3)

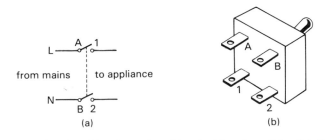

(a)   (b)

Figure 79.3   (a) Switching the mains with a double pole switch   (b) Rear view of a typical double pole switch

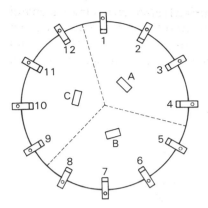

Figure 79.5 Rear view of a 3 pole/4 way rotary switch

A relay uses the electromagnetic effect. You can see the circular coil on the right. The contacts are on the left.

Two-pole/two-way switch (or DPDT)

As the name suggests, this is two separate two-way switches operated by one lever. It is sometimes called a **changeover** switch. One use for this is in an intercom where each speaker can be used either as a speaker or a microphone. See Figure 79.4.

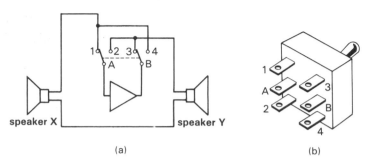

Figure 79.4 (a) An intercom uses a double pole/double throw switch. In the position shown speaker X is connected to the input and Y is connected to the output (b) Rear view of a typical double pole/double throw switch.

Multi-pole/multi-way switches

A common arrangement is **3 pole/4 way**. These are generally **rotary-type** switches which are operated by turning a spindle. A typical example is shown in Figure 79.5.

Relays

A **relay** is a switch which is operated by sending a small current through a coil. This attracts the lever which then moves the contacts. See Figure 79.6. The contacts may be DPDT or SPDT, etc.

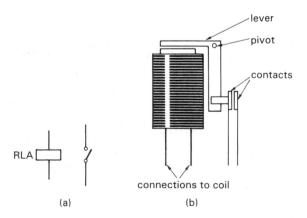

Figure 79.6 Relays (a) Symbol (b) Main parts of a typical device

The contacts can have a large **current rating** so a relay is a useful way of switching a large current on or off (at the contacts) using a small current (in the coil). A relay is often used at the output of an electronic switch so that large, mains equipment can be switched on and off. (Figure 79.7)

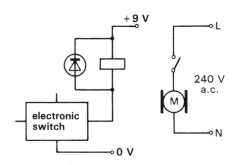

Figure 79.7 Controlling a mains-operated motor by using a relay

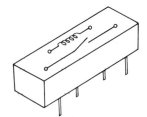

Figure 79.8 A typical reed relay

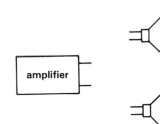

Figure 79.9

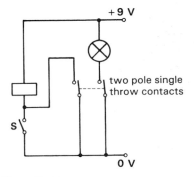

Figure 79.10

Reed relay

A **reed relay** is a smaller version. The contacts and coil are usually hidden inside a plastic case. The contacts are smaller and can handle less current than the electromagnetic relay. Their advantages are that the coil needs less current to operate and they last longer as well as taking up less space. (Figure 79.8)

Circuit breakers

These are safety devices which switch off the current automatically if a fault occurs. Some of them operate if the mains current exceeds a certain level. Other types are triggered when a current flows in the earth wire (perhaps because of a 'short'). They can either be electromagnetic devices similar to relays or thermal devices operated by a bimetallic strip.

Questions

1 What do the following abbreviations stand for? (a) SPDT (b) DPDT.
2 Copy the block diagram in Figure 79.9 and put in a switch which will allow us to select which of the two speakers are used.
3 (a) Why must a relay be used if we want a transistor switch to control a 240 V a.c. motor?
   (b) What precaution must we take when using relays in transistor circuits?
4 Look at the circuit diagram in Figure 79.10 and explain what will happen if the switch S is closed and then opened again. Can you think of a use for this?

80

Calculations

This chapter goes through some of the common calculations that sometimes have to be done in electronics.

Resistance, voltage and current

The easiest way of remembering the formula is to use the **magic triangle** shown in Figure 80.1. If we cover up the one we want to calculate, we are left with the formula to use. Consider the circuit in Figure 80.2.

To find **resistance**, covering up R in the triangle leaves V/I so:

$$R = V/I$$
$$= 6\,V/3\,mA$$
$$R = 2\,k\Omega$$

(remember if I is in mA then R is in kΩ)

To find **current**, covering up I in the triangle leaves V/R so:

$$I = V/R$$
$$= 6\,V/2\,k\Omega$$
$$I = 3\,mA$$

To find **voltage** (or **p.d.**), covering up V in the triangle leaves $I \times R$ so:

$$V = I \times R$$
$$= 3\,mA \times 2\,k\Omega$$
$$V = 6\,V$$

Figure 80.1 The 'magic triangle' for calculating p.d., resistance and current

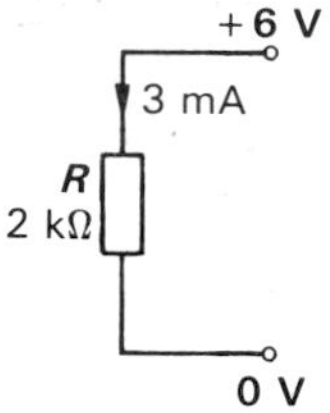

Figure 80.2 Resistor connected across a p.d.

Resistors in parallel

To find the effective resistance of two resistors in parallel (see Figure 80.3) use the formula:

$$R = \frac{R_1 \times R_2}{R_1 + R_2}$$

For example, if $R_1 = 22\,k\Omega$ and $R_2 = 10\,k\Omega$, the effective resistance (R) is

$$\frac{22 \times 10}{22 + 10}$$

$$= \frac{220}{32}$$

By calculator, this gives $R = 6.875\,k\Omega$. Always round this up to two figures, so $R = 6.9\,k\Omega$.

Useful check: the answer should always be **less** than either of the two resistors.

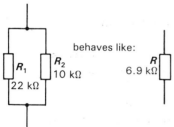

Figure 80.3 Resistors in parallel

Power

Look again at Figure 80.2. The **power** lost (or rate at which heat energy is produced) in a resistor is best calculated by either of these two formulae:

(1) $P = I^2 \times R$
       $P = 3^2 \times 2$ mW
       $P = 9 \times 2$ mW

(2) $P = V^2/R$
       $P = 6^2/2$ mW
       $P = 36/2$ mW

Either way, $P = 18$ mW.

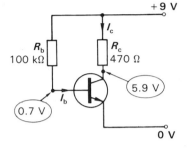

+9 V

R_b
100 kΩ

R_c
470 Ω

I_c

5.9 V

I_b

0.7 V

0 V

Figure 80.4   Simple biassing circuit for a transistor

Transistor calculations

A common problem is based on the circuit in Figure 80.4. Suppose that the **current gain** of the transistor is 80.

Calculating p.d. across R_b:
- assume that the voltage at the base is 0.7 V
- the p.d. across R_b must be 8.3 V (9 V − 0.7 V).

Calculating the current flowing into the base through R_b:
- since the p.d. across R_b is 8.3 V, then
- the current flowing through R_b is 8.3 V/100 kΩ, so
- current flowing = 8.3/100 mA
              = 0.083 mA

Calculating current flowing into collector:
- since the current gain is 80, then
- the collector current = 80 × base current
- collector current = 6.64 mA (80 × 0.083 by calculator)
- round this up to two figures: I_c = 6.6 mA

Calculating p.d. across R_b:
- the collector current must flow through R_c, so
- current through R_c = 6.6 mA
- the p.d. across $R_c = I \times R = 6.6 \times 0.47$ (keep R in kΩ), so
- p.d. across R_c = 3.1 V (by calculator and rounded up)

Calculating voltage at the collector:
- if supply voltage is 9 V and p.d. across R_c = 3.1 V, then
- voltage at collector = 5.9 V (9 − 3.1)

   Question: what if the p.d. across R_c works out to be more than the supply voltage? **Answer**: the transistor is turned fully on and the voltage at the collector is 0 V.
   Note: this implies that the collector current must be less than the calculated value. We can only use '$I_c = h_{FE} \times I_b$' if the transistor is not fully turned on.

81
Circuit Symbols

Resistors

fixed value

variable

variable
potential divider

old resistor symbol

thermistor

thermistor
(alternative symbol)

LDR

Capacitors

fixed value

variable

electrolytic

electrolytic
(alternative symbol)

Inductors

fixed value

variable

transformer

Diodes
(The circular envelope in these diode symbols may be omitted.)

diode

zener diode

silicon-controlled
rectifier

diode (old symbol)

Semiconductors

n-p-n transistor

p-n-p transistor

field-effect transistor (FET)

Indicators

lamp

neon

LED

Connections

wire

fuse

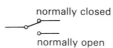
wires crossing

earth

wires connecting

terminal

Switches

push to open

push to close

toggle switch

relay coil

normally closed

normally open
relay contacts

reed relay

Input and output devices

microphone

tape head

signal generator

pick-up

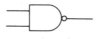
speaker

earphone

bell

buzzer

motor

Gates and other circuits

OR

NOR

XOR

AND

NAND

NOT or INVERTER

operational amplifier

low pass filter

high pass filter

1 Here are some electronic component symbols:

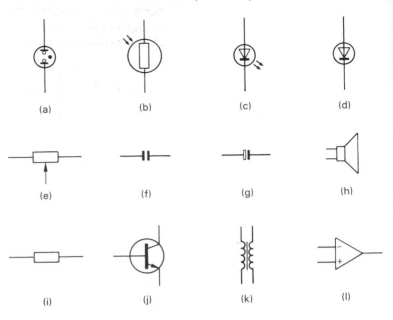

(a)   (b)   (c)   (d)

(e)   (f)   (g)   (h)

(i)   (j)   (k)   (l)

Which component or components (when working correctly):
(a) can, by themselves, increase an alternating e.m.f.,
(b) must be connected with one lead positive and the other negative,
(c) can be used as an input device or sensor,
(d) can store an electric charge
(e) only work with a high voltage,
(f) can be made from carbon,
(g) can be made from silver and mica,
(h) have an $I_c\,max$ rating,
(i) produce light,
(j) have a VA rating,
(k) always have an extremely high resistance,
(l) can be 'LIN' or 'LOG' types,
(m) work by the electromagnetic effect,
(n) can be used as an output device,
(o) can be used in a tuned circuit?

2 Draw the circuit symbols for the following:
(a) a Zener diode
(b) a DPDT (or 'changeover') switch
(c) a variable capacitor
(d) a p-n-p transistor
(e) an SCR.

3 A manufacturer suggests that his thermistor can be used in constructing temperature sensing devices by using it to control the lighting of coloured light-emitting diodes (LED). One way is to use a transistor as in the circuit shown in Figure 82.1:

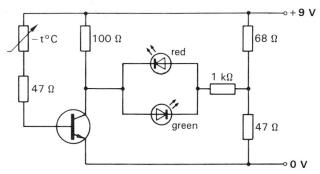

Figure 82.1

(When calculating the currents flowing in this circuit there is no need to take account of the currents carried by the LEDs.)

The thermistor changes its resistance with a change of temperature in this way:

temperature in °C:	−30	−15	0
resistance in kΩ:	30	10	3

(a) With the thermistor maintained at a temperature of −15°C calculate: (i) the base current flowing in the transistor; (ii) the collector current, the transistor having an $h_{FE} = 120$; (iii) the voltage at the transistor collector, at the junction with the two LEDs.
(b) State which of the LEDs will be alight under these conditions and explain why.
(c) In the circuit in Figure 82.1 the thermistor temperature can now be changed to one of the other values listed in the table; −30°C or 0°C. Do calculations, similar to those done in part (a), to explain why this LED is alight at the temperature you have selected.
(d) At some temperatures it is found that neither LED is alight. Explain how and why this happens. (*AEB 84 paper 1*)

4 Figure 82.2 shows three devices that are each connected in turn between points A and B of the potential divider in Figure 82.3.

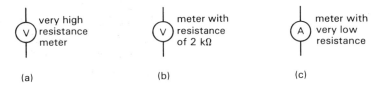

(a)   (b)   (c)

Figure 82.2

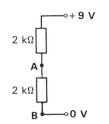

Figure 82.3

Describe and explain exactly what you would see indicated by each of the three devices.

Index